O9-AHU-053

Thank You for VOTING

Thank You for VOTING

THE MADDENING, ENLIGHTENING, INSPIRING TRUTH ABOUT VOTING IN AMERICA

ERIN GEIGER SMITH

HARPER
An Imprint of HarperCollins*Publishers*

 For information, address HarperCollins Publishers, 195 Broadway, New York, NY 10007.

HarperCollins books may be purchased for educational, business, or sales promotional use. For information, please email the Special Markets Department at SPsales@harpercollins.com.

FIRST EDITION

Library of Congress Cataloging-in-Publication Data has been applied for.

ISBN 978-0-06-293482-6

20 21 22 23 24 LSC 10 9 8 7 6 5 4 3 2 1

For Bryan

For Mom

CONTENTS

Preface ix

PART ONE: HOW WE GOT THE VOTE

CHAPTER ONE Democracy in Name Only 3

CHAPTER TWO Long-Suffering for Women's Suffrage 23

CHAPTER THREE Voting Problems and Voting Solutions 45

PART TWO: HOW TO GET PEOPLE TO VOTE

CHAPTER FOUR Transforming Non-Voters into Voters 73

CHAPTER FIVE Making Voting Their Business 95

CHAPTER SIX Thank You for Voting 109

PART THREE: KNOW BEFORE YOU VOTE

CHAPTER SEVEN Gerrymandering: Over the Line? 127

CHAPTER EIGHT Knowing the News Is Real 145

CHAPTER NINE Understanding Polling 163

CHAPTER TEN Explaining the Electoral College 175

Thank You for Voting: A Checklist 193
Thank You for Voting: Tell Your Friends 197
Acknowledgments 199
Notes 203
Index 227

PREFACE

League of Women Voters "Get Out the Vote" Wagon, Anne Arundel County, Maryland, 1940s (Courtesy of League of Women Voters)

Nobody will ever deprive the American people of the right to vote except the American people themselves—and the only way they could do that is by not voting."

President Franklin D. Roosevelt's quote resurfaces every election season to encourage voter participation. It conjures an image of an American flag waving gracefully against a cloudless sky and reminds us of our personal power to steer the country on health care, education, the economy, climate change, and everything else that matters. Voting is one of America's greatest sources of pride. It also exposes our greatest contradictions.

The United States has a voting problem: not enough of us do it.

To solve that problem, and it is solvable, we need to better understand our voting history and how it affects turnout today. Making that connection allows an honest evaluation of flaws in our electoral system. But it also means celebrating creative and effective get-out-the-vote efforts, highlighting the importance of adding fun and friendship to the process, and learning about solutions that make it easier to register and vote. Taking time to break down topics that stump even the most educated voters will empower us to enter the voting booth confident in our choices all the way to the bottom of the ballot.

The goal of this book is to provide the information necessary for you to vote, to convince you of the importance of voting in every election, and to encourage you to recruit the people you're closest with to join you. (Yes, social-media-close counts as close.)

Roosevelt's quote is the Instagram version of the country's more nuanced reality. The longest-serving president delivered the soundbite in a 1944 campaign-season address from the White House, at a time when states deprived many American citizens of the right to vote and had just years earlier barred many more.

The same year as Roosevelt's speech, the U.S. Supreme Court ruled against the Texas "white primary," which kept black people from voting. True access to the polls for African Americans in the South was more than twenty years away. Many Native Americans had more than a decade ahead of them in their struggle for voting rights; some Asian immigrants were still permanently barred from voting. Women had voted nationwide for less than twenty-five years.

Roosevelt knew all of this. He began his speech on that patriotic note but soon acknowledged all the ways this initial image of America was a flattering filter. "The right to vote must be open to our citizens irrespective of race, color or creed . . . The sooner we get to that basis of political equality, the better it will be for the country as a whole."

Roosevelt wanted to promote our ideals about voting and bury, even if just for a moment, discussion of the work still needed to approach political equality. Ignoring this bigger truth is a habit Americans have developed into an art form.

After the most recent presidential election and the country's contrasting reactions of jubilation and despair, I couldn't stop thinking about the power in voting, and what it means in a country that appears so divided on cornerstones of democracy—the First Amendment, the rule of law, immigration, whether watching eighteen hours a day of cable news is a good idea.

I live in New York City and am a journalist. I've written articles on entrepreneurs at the top of their game, tech executives trying to survive Supreme Court battles, artists creating postcards with political messages, and the very real struggle to complete *Infinite Jest.* But my own story begins in Liberty, Texas, its population now just shy of ten thousand people. It was an even smaller town when I lived there. Towering oak trees shade the three-story limestone courthouse in a quintessential town square. It's Liberty's seventh courthouse, the first built of hewn logs when Texas was still part of Mexico. The town burger joint is across the street, and multiple churches are within walking distance, their steeples among the tallest things in town.

I feel a strong connection to both Manhattan and my hometown, but around election time they seem like different universes. Liberty County is overwhelmingly Republican, while my current neighborhood votes heavily in favor of Democrats. At the end of

2016, it felt like New York and other major cities had long lived with the idea that their opinions were the only ones that mattered, while Liberty had finally banded together with its like-minded small towns to turn an annoyed whisper into a thunderous shout.

The two places' very different motivations for supporting candidates are clear to me. And while our fear that we won't be represented, that our wants and needs will be ignored if our candidate doesn't win, is emblematic of how divided we are, it also highlights the importance of electing politicians willing to consider the concerns of constituents who voted for somebody else.

Despite the constant political noise about our strong feelings, our hopes, and our abiding deference to the idea of majority rule, so many of us are silent come Election Day. Even with a recent uptick in turnout in 2018, those who do vote can't pick up the slack for their peers who don't.

Each generation votes at lower rates than the one older, and at times about a quarter of young adults speak for all of them. The same downhill slide holds when it comes to education levels. Adults with postgraduate degrees love to vote—around 80 percent turnout for presidential elections. But then the drop starts and doesn't stop, with Americans who didn't graduate high school sometimes voting at rates two times lower. There are racial disparities in who votes, too. While African Americans and white people vote in similar numbers in presidential elections, a significantly lower percentage of Hispanics and Asian Americans participate.

Each person who does vote is motivated by his or her own unique combination of issues and histories. Loyalty to a political party gets some to the polls; others vote when their union tells them to, or when their favorite TV commentator does. The most predictable voters do it out of habit, driven by civic duty as much as by who is on the ballot. Every election offers young people the thrill of voting for the first time. Some people cast a ballot to honor grandparents who never had the privilege.

A friend recalls adults in Sunday school who said that finding candidates who would outlaw abortion was a top priority. An Ohio autoworker's vote in the last presidential election was earned by the person he thought could save the livelihood of his struggling town. Those despairing over the environment are interested only in vocal proponents of greener policies. "It's the economy, stupid" has become a political trope, but impact on one's own wallet is the driving factor for many.

The Supreme Court. The Second Amendment. Minimum wage. The opioid crisis. The motivations are countless, but they all result in one act: voting. If others who share your views stay home, your voice isn't optimized. We'll never know what the country truly wants and needs unless more of us participate.

This book is divided into three sections. The first explains how and when different groups of Americans—African Americans, women, and young people, among others—got the right to vote. It also looks at current examples of voter suppression, and methods some states use to increase registration and turnout. The second section takes you inside innovative get-out-the-vote movements to demonstrate how their participants' actions and attitudes can be replicated to boost turnout in your own peer or work group. And the final section provides straightforward explanations of vexing voting topics—gerrymandering, political news and polling, and the Electoral College. The last pages feature checklists to help you prepare to vote and spread the word.

I spent more than a year researching the history of voting in America, filling two-toned blue file folders labeled with topics like "Youth Vote" and "Native American Voting." The "Fun Facts" file included the tidbit that the United States has the busiest election calendar in the world; Americans have more opportunities to vote in ten years than Japanese citizens have in a lifetime. I also came across many distressing facts: for instance, the number of Americans who were eligible to vote but who didn't cast a ballot in the 2016 presidential election is greater than the number who voted for any one candidate. The country showed rousing support for President Oops, Didn't Choose One.

I was awed by the jaw-droppingly impressive writings of early feminists. A book on congressional maps titled *Ratf**ked* was a true page-turner. I also came across simple but joyful stories and images: videos of college students celebrating registering to vote, jeans with "Vote" embroidered in script on a rolled-up cuff, and a story about early 1900s suffrage dolls made of cotton purchased to support U.S. farmers during World War I.

Join me in learning how the course of women's rights was changed by a young legislator who defied his constituents and listened to his mother, and how the chief justice of the United States wrote an opinion diminishing a voting rights law thirty years after he'd questioned it as a twentysomething. You'll meet Yara Shahidi, a television phenom inspiring a youth voting revolution, and everyday high schoolers doing the same. Interviews with experts and reporters provide tips on how to read political news and polls without going insane. (Okay, polls may still drive you crazy, but you'll understand them better.) Historians explain how it almost never happens that the popular vote and the Electoral College count turn out differently. Except, of course, when they do.

In every chapter are examples of individuals who decided to get involved in increasing voter turnout and catapulted in. There are righteous single ladies, mothers of young babies, and intrepid former slaves who collectively spent seventy years securing the right to vote. Modern voting heroes include a twentysomething whose advocacy got her state constitution amended and a thirtysomething whose campaign to promote voting took Hollywood by storm. Then there's the fortysomething who built a coalition of four-hundred-plus businesses that gave employees time off to vote.

I want you to close this book with a desire to vote every chance you have, in every race on the ballot. Some of the most important decisions that affect your life are made in years when there isn't a presidential election. Governors and secretaries of state and state legislators wield an amazing amount of power. As the coronavirus spread across the country in early 2020, mayors and city council members made vital, difficult decisions on how to keep essentials available and protect citizens' lives.

The truth is no one knows for certain what makes turnout for a given candidate a sure bet. But everyone who studies voting agrees that the person with the best shot at getting particular people to vote is YOU!

Columbia University professor Donald Green, one of the top thinkers on what motivates voters, explained the most effective turnout tactic: getting your group of politically minded friends, regardless of party affiliation, to promise to vote. It's as simple as "You pledged to vote, so did I. And now we're accountable," Green said. It's what has the best chance of working on your non–politically minded friends, too.

What I'm not doing is directing your vote in any way. I believe the more people who vote, the more representative and responsive our government will be. Figuring out the best way to achieve the largest turnout should have nothing to do with one side or the other and everything to do with supporting democracy. This book is nonpartisan, but it is staunchly pro-voting. I refuse to accept that access to the polls and promoting turnout is a partisan issue, and I hope you feel the same.

Roosevelt's campaign speech was a plea for voter turnout, and its lines resonate seventy-five years later. He said that we shouldn't "be slackers on registration day or Election Day," and that only a large outpouring of voters could definitively show who the "masses of the American people" wanted to win. The "continuing health and vigor of our democratic system depends on the public spirit and devotion of its citizens." We need citizens expressing themselves at the ballot box.

He's right. Roosevelt may have started with a too-rosy outlook on voting in America. But, as always, we're striving to get better. So let's listen to him. Let's not be slackers, let's get out the masses, and let's devote ourselves to democracy. I hope you'll vote. And that you'll bring your family and friends, your neighbors and coworkers, your competitors and rivals. You get the point.

Thank you for voting.

PART ONE

How We Got THE VOTE

CHAPTER ONE

DEMOCRACY IN NAME ONLY

Protesters outside the White House in Washington, DC, March 12, 1965 (Warren K. Leffler/Courtesy of the Library of Congress)

The dates of voting rights victories can sound like ancient history, but the 1965 Voting Rights Act gave many people alive today their first opportunity to vote, and other groundbreaking voting laws benefited the parents and grandparents of today's Americans. A white woman born in 1900 would have been among the first able to vote nationwide as soon as she turned twenty-one. Many immigrants of Asian descent born that same year wouldn't have their citizenship approved until the year they turned fifty-two.

An African American born at the turn of the twentieth century and living in the South may not have cast a ballot on Election Day until she was sixty-five years old.

Election Day in the modern world often ends with anchors on the various news channels breaking the country down to our most basic facts. CNN's John King is famous for standing at his giant touch screen of the United States, employing his savant-like electoral knowledge to zoom in on states and congressional districts to discuss demographics, party preferences, and population numbers. Those of us who can't turn away from the political drama spend the night in front of the TV, logging onto our most trusted election polling website or tossing off proclamations to family and friends or on Twitter. After voting, in other words, there's nothing to do but wait.

But in the early days of voting in America, Election Day brought a party-like atmosphere, with taverns often serving as voting locations. There was drinking and laughing and, with no secret ballots or rules about politicking close to the polls, plenty of yelling and in-your-face voter intimidation, sometimes violent. Of course, if you weren't a landowning white man over twenty-one, well . . . you weren't so welcome. An early American John King wasn't needed to break down the characteristics of the electorate, because they were simple: they mostly looked like the founding fathers, and not the versions onstage in Lin-Manuel Miranda's *Hamilton*.

It took the United States a very long time to get to an era in which men and women of all ethnicities, races, religions, and income levels were able to vote in all elections. Those of us lucky enough to cast a ballot today with relative ease and no disturbance often forget what it took to get here. Those of us for whom voting remains a struggle likely still feel the weight of history.

Our country now recognizes that citizens eighteen and over have a right to vote, assuming they've met registration requirements and (in some states) haven't committed a felony. But in America's infancy, the British voting requirements of male and landowning were largely carried over. Though the new country was asserting its independence and seeking to build a democracy, those in power didn't see reasons to expand the electorate. They didn't present extensive arguments explaining why such an expansion was impossible; they simply preferred to protect the benefits of their privilege and power. Limiting the vote for reasons of self-interest is a scenario that played out repeatedly as the country grew.

In 1776, founding father John Adams wrote about what would happen if men who had no property were allowed a voting voice.

> *New claims will arise. Women will demand a vote. Lads from twelve to twenty-one will think their rights not enough attended to, and every man who has not a farthing, will demand an equal voice with any other in all acts of state. It tends to confound and destroy all distinctions, and prostrate all ranks, to one common level.*

That quote is cleaned up a bit to reflect current punctuation and capitalization conventions, and here's my shorter interpretation of Adams: We must keep the voting power for ourselves, or people will see that they too are entitled to it in a true democracy. Everyone would be equal! The nightmare!

Even then, though, there were exceptions to the white-men-only rule. In 1776, Pennsylvania's new constitution did not require property ownership, so freemen who paid at least some taxes could vote. The next year, Vermont's constitution abolished slavery for men over twenty-one and gave those males the right to vote.

For about thirty years, starting in 1776, property-owning women and free black men in New Jersey could vote. That ended in 1807, when the state legislature limited voting to white property-owning men. Additional regression for African Americans followed. From

1819 forward, any state admitted to the Union prohibited black people from voting, and multiple states either made it harder for black men to vote or outlawed it completely. Pennsylvania, for instance, limited voting to whites only in 1838, and did so despite the reasoned and heartfelt argument in an "Appeal of Forty Thousand Citizens, Threatened with Disfranchisement," written by Robert Purvis, the first African American member of the Pennsylvania Abolition Society. "When you have taken from an individual his right to vote, you have made the government, in regard to him, a mere despotism; and you have taken a step towards making it a despotism to all," it read in part.

Purvis pointed out that free black men had been voting in Pennsylvania for a half century, and that raises a question we should all consider: What happened when women and black people could vote? The sky did not fall. The country survived and grew. Lawmakers were presented with the concept of broad forms of suffrage from the start; the majority of them just didn't want it.

Adams's prediction that all people would eventually demand the benefits of citizenship was correct. Americans learned to use their voices to advocate for themselves, and used the Constitution's language to argue that equal should mean equal and that citizenship didn't count for much if you didn't have the vote. As the nation grew, women, African Americans, Native Americans, and immigrants were among the groups who would have to overcome blatant discrimination, skeptical politicians, physical attacks, and court battles to secure their right to vote.

Those seeking access to the polls were never completely welcomed with open arms—the fights were long and ugly, and disenfranchised groups were sometimes pitted against one another. And while we now look upon voting victories won—the Fifteenth Amendment (granting men of all races the right to vote), the Nineteenth Amendment (securing women's suffrage), and the Voting Rights Act (ensuring minority groups could *actually* vote)—as achievements of a true democracy, they certainly weren't met with countrywide celebration. Chapter Three lays out the ways in which minority groups still sometimes face barriers that make voting difficult.

In his definitive *The Right to Vote*, Harvard Kennedy School professor of history and social policy Alexander Keyssar writes of the fits and starts of voting rights in America:

> *The evolution of democracy rarely followed a straight path, and it always has been accompanied by profound antidemocratic countercurrents. The history of suffrage in the United States is a history of both expansion and contraction, of inclusion and exclusion, of shifts in direction and momentum at different places and at different times.*

My version of Keyssar: When the powerful were finally forced to giveth, they immediately made backroom plans to taketh away.

African Americans and the Vote

There is no sugarcoating the treatment of black people in America's voting history. Individuals and politicians at every level of government repeatedly and systematically discriminated against African Americans, and denying their right to vote has been one of the pillars of that prejudice.

The U.S. Supreme Court decided the infamous *Dred Scott* case in 1857, less than five years before the Civil War, ruling that black people could not be citizens of the country, no matter if they were slaves or free. When the war began, only a handful of states' laws allowed black men to vote equal to whites. But black men fought in the Civil War in great numbers, and began pushing for equality in all states, including for voting rights, as the war came to an end.

In 1865, a group of fifty-nine African Americans from Tennessee presented a petition to a pro-Union convention in Nashville, asking for the right to vote and referring to those African Americans still fighting valiantly in the war. They acknowledged that many would find the idea of black people voting shocking at first, but reminded the audience that public opinion among white people had once been that black people wouldn't be good soldiers, and that free black men had voted in Tennessee earlier in the century. That same year, black citizens of New Orleans organized a mock election to show how seriously they took the right to vote. Around the same time, freedmen in North Carolina formed the Equal Rights League.

The Fourteenth Amendment, passed in 1866 and ratified in 1868, granted citizenship to all people born or naturalized in the United States, and guaranteed all citizens "equal protection of the laws." It also aimed to force Southern states to allow black men to vote, saying that if of-age males weren't allowed to vote, the discriminating state's representation in Congress would be reduced in proportion to the number of people improperly kept from voting.

The Reconstruction Acts of 1867–1868 had even stronger language, requiring that in order to be readmitted to the Union, Southern states had to approve the Fourteenth Amendment and draft state constitutions that allowed black males the right to vote. Seven states had been readmitted to the Union by July 1868. "Black enthusiasm for political participation was so great that freedmen often put down their tools and ceased working when elections or conventions were being held," Keyssar writes.

Reconstruction, which lasted for about a decade, was an early high point for African American representation. "Black political power was strongest in South Carolina," historians wrote in *The Atlantic*; its state constitutional convention in 1868 was majority black, as was one house of its state legislature from then until 1876.

In 1870, over the protest of other elected officials, Hiram Revels took his oath of office as a senator for Mississippi; Joseph Rainey of South Carolina became the first African American member of the U.S. House of Representatives that same year. During this time period, an estimated two thousand black men held elective office. (Mississippi has not elected a black U.S. senator since Reconstruction. The state's population is currently about 38 percent black.)

Still, there was great opposition to upward mobility for African Americans. The Ku Klux Klan and other vigilantes began their reign of terror, threatening, injuring, and killing freedmen. Support for African Americans' rights in the North was uneven. Those supporting nationwide voter protection for black people argued that a federal suffrage amendment was a necessity. Ratified in 1870, the Fifteenth Amendment said that the right to vote "shall not be denied or abridged by the United States or by any State on account of race, color, or previous condition of servitude."

The language of the Fifteenth Amendment was heavily debated, and narrower in the end than what lawmakers in favor of universal suffrage wanted. Some proposals included language that would have outlawed the exact types of discrimination that would hinder voting rights for decades. "What opponents of a broad amendment rejected in the end was the abolition of discrimination based on nativity, religion, property, and education. They wanted to retain the power to limit the political participation of the Irish and Chinese, Native Americans, and the increasingly visible clusters of illiterate and semi-literate workers massing in the nation's cities," Keyssar writes.

The end of the Reconstruction era came shortly after the controversial presidential election of 1876. Republican candidate Rutherford B. Hayes lost the popular vote, but a compromise with Southern Democrats over disputed Electoral College votes gave Hayes the presidency. The compromise also required the federal government to pull its troops out of the South, giving politicians determined to roll back the rights of black Americans the opportunity to do so. Jim Crow laws that required race-specific designations like separate seating for blacks and whites and extreme voting limitations flourished.

It's astonishing how much evil creativity people put into suppressing black men's votes. A Florida provision enacted in the late 1880s, for example, had separate ballots for different political races and required voters to place the ballot for each race in the correct box or their vote wouldn't count. But those who couldn't read the box labels had little chance of getting it right. Over the next several years, African American voter turnout dropped from 62 percent to 11 percent.

The systematic methods enacted in the South to disenfranchise blacks were successful. In 1896, about 130,000 black residents were registered to vote in Louisiana; two years after the state enacted its 1888 "Grandfather Clause," only about 5,300 black men voted. By 1904, only 1,300 African Americans were registered. The illiterate or non-property-owners could register to vote only if their fathers or grandfathers had been legally able to vote in 1867. These grandfather clauses kept descendants of slaves from the polls.

The country was—there is no other way to put this—divided over whether African Americans should be treated as equals. Republican Benjamin Harrison made protecting black men's voting rights a part of his successful 1888 presidential campaign. Harrison spoke out against the disenfranchisement of blacks in the South, saying he wouldn't "purchase the presidency by a compact of silence." Hoping to capitalize on Harrison's win, lawmakers introduced a bill to allow a small group of a state's citizens to petition a federal judge to appoint supervisors to oversee the state's federal elections. Among the sponsors of the bill was Massachusetts representative Henry Cabot Lodge. The bill's provisions included monitoring registration lists, ensuring that non-citizens did not vote, and certifying vote counts. It passed in the House but not the Senate.

By 1940, only 3 percent of eligible African American Southerners were registered to vote, due to Jim Crow laws like literacy tests and poll taxes. It wasn't until passage of the Voting Rights Act in 1965 that African Americans' right to vote was protected nationwide in any real way. The VRA transformed voting in the South by enacting protections similar to those the Lodge bill had proposed seventy-five years before.

WHO WAS JIM CROW?

We know that Jim Crow laws enforced racial segregation and discriminated against African Americans: there were separate water fountains for black people, and poll taxes were levied—a fee designed to make voting prohibitive. The term "Jim Crow" is used to this day to describe laws with racist undertones. But who was Jim Crow?

He wasn't a real person but a character, played by a white actor named Thomas Dartmouth "Daddy" Rice in the 1830s. Rice wore blackface and shabby clothes, used an offensive dialect attributed to slaves, and portrayed the character as slow-witted. He is said to have based it on a black man (some sources say a child) he saw singing a song called "Jump Jim Crow" in Louisville, Kentucky. The act was a hit.

"Jumping Jim Crow" and "Jim Crow" "became shorthand . . . for describing African Americans," Eric Lott, a professor at the City University of New York and author of a book on minstrel performances, told *National Geographic*. It's unclear exactly how the name came to describe the laws.

As early as 1887, the *New York Times* used it as a description for the train car in Georgia a black man complained of being forced to sit in. A *Times* opinion piece in 1915, headlined "Breaking the Jim Crow Law," detailed how black people in Tennessee were frustrated that white men were taking seats in black train cars when there weren't seats available in their own: the *Chattanooga Times* "justly and severely remarks that these men seem to hold the doctrine . . . that the black man has no rights the white man is bound to respect." It wouldn't be until fifty years after that piece that the Voting Rights Act began to loosen Jim Crow's hold on America.

You might wonder what the U.S. Supreme Court was up to during the ninety-five years between passage of the Fifteenth Amendment and the Voting Rights Act. What did they think about all these laws created specifically to get around the Fifteenth Amendment?

In 1896, *Plessy v. Ferguson* established the doctrine of "separate but equal." In the early 1880s, state legislatures began passing laws requiring separate train cars for black passengers and white passengers. In 1892, a black man named Homer Plessy, working with the New Orleans Citizens' Committee, decided to challenge the law. He boarded the "Whites Only" car on a Louisiana train and refused when he was asked to switch to the area designated for black people. He was arrested and the lawsuit followed. I studied the *Plessy* opinion in law school, but still found it shocking when reading the language again recently. In response to the argument that separate facilities themselves indicated inequality, the Supreme Court majority opinion said that "if this be so, it is not by reason of anything found in the act, but solely because the colored race chooses to put that construction upon it." (Translation: if they feel that way, it's on them.)

Justice John Marshall Harlan criticized his fellow justices for being dishonest about the purpose of the legislation, writing in a stinging dissent that it was obvious to all that the law was not designed to treat blacks and whites equally. It was designed to keep black people out of cars designated for white people. (Black people could be in "white" cars when they were "nurses" accompanying white children.) Justice Harlan likened the decision to *Dred Scott* and correctly predicted that the decision would be harmful to the country. It is widely considered to be a stain on the Supreme Court's history.

Two years after *Plessy*, the Supreme Court upheld Mississippi's literacy test as constitutional, and nearly forty years later, in 1937, ruled that Georgia could continue to require voters to pay a poll tax. *Plessy* would eventually be overruled by *Brown v. Board of Education* (1954), which ended "separate but equal" by calling for the end of segregated public schools. But as late as 1959, the court ruled that North Carolina's literacy tests did not violate the Fourteenth and Fifteenth Amendments.

The 1965 Voting Rights Act was the undoing of discriminatory voting laws, but the VRA might never have happened without the country's bearing witness to a violent day in Alabama.

Even though the Student Nonviolent Coordinating Committee spent two years working to register African Americans to vote in and around Selma, Alabama, only 335 of the county's 15,000 citizens eligible to register were registered in 1965. A march from Selma to Montgomery, Alabama, was planned for March 7 to protest the shooting death by a state trooper of twenty-six-year-old civil rights activist Jimmie Lee Jackson.

John Lewis is one of the longest-serving U.S congressmen and a civil rights legend, but that morning he was a twenty-five-year-old activist carrying a backpack with fruit, a toothbrush, and two books. (Lewis would weep in 2019 when discussing the 1867 voter registration card of his great-great-grandfather. His colleague noted that until Lewis walked at Selma and helped achieve the civil rights changes that followed, members of Lewis's family had not been able to vote since Reconstruction.)

A police officer fractured Lewis's skull with a club shortly after the march began, and Lewis was only one of many injured in what came to be known as Bloody Sunday. Images of the incident were publicly televised, the nation finally reacted in horror, and President Lyndon Johnson acted quickly to try to right the decades of wrongs that kept black people from voting in the South. Johnson had reportedly told his attorney general, in a phrase easy to imagine in a Texas drawl, to write the "god-damnedest toughest" voting rights laws. He saw the opening to pass them.

Just eight days after Bloody Sunday, Johnson addressed a joint session of Congress. "Many of the issues of civil rights are very complex and most difficult," he said. "But about this there can and should be no argument. Every American citizen must have an equal right to vote."

Johnson's focus on the importance of the vote, writes journalist Ari Berman in *Give Us the Ballot*, "brilliantly framed the cause of voting rights not as an issue of black versus white but as right versus wrong. Despite the country's tortured racial history, the president argued that denying the right to vote undermined the ideals of liberty and freedom that made America exceptional." (Berman has written extensively on the history of voting rights and on suppression of voting rights today, most recently for the liberal magazine *Mother Jones*.)

The Voting Rights Act, signed into law in August 1965, reinforced the Fifteenth Amendment's requirement that the vote cannot be denied on the basis of race. Along with subsequent amendments, it outlawed tactics often used to keep black people from voting, and required states and counties with a history of denying non-whites the vote to submit for approval by the federal government any planned changes to their voting laws.

The law's impact was swift and real—by the end of the year, about 250,000 new black voters had registered, and by the end of 1966, 50 percent or more of eligible African Americans in nine of thirteen Southern states were registered. By 1967, African American registration in Mississippi had catapulted from just 6.7 percent before the VRA to 59.8 percent after.

The Voting Rights Act had expiration dates, both as a concession to lawmakers who opposed it and to allow regular evaluations to determine if it was still useful. If states in which there was discrimination cleaned up their acts and the VRA wasn't needed anymore, it could expire. In fact, however, it was renewed, strengthened, and extended multiple times by presidents and lawmakers of both parties, including into the Obama administration.

In the years after its passage, the Supreme Court would repeatedly uphold or clarify the VRA's reach and rule against laws believed to be created to stifle the minority vote. The court struck down New York's English-language literacy test in 1966; it was said to focus on disqualifying Puerto Rican natives from voting.

The Voting Rights Act made access to the polls the true and enforceable law of the land; it allowed citizens who believed that their voting rights were being violated to bring challenges in court. But just as with voting rights progress before it, there were sustained and repeated efforts to roll the VRA back and make it less effective. One such effort included crafting ways of electing local officials that would dilute the minority vote for decades.

The Fourteenth and Fifteenth Amendments and the Voting Rights Act are used to this day to fight tactics that disproportionately affect minority voters, including the closing of polling locations, improper removal of voters from the voting rolls, and the drawing of state and federal lines for election districts.

Native Americans and the Vote

It was the Voting Rights Act that also fully extended the right to vote to Native Americans. The Fifteenth Amendment did not speak to discrimination based on where people were from, their religion, economic level, or education. States continued to deny Native Americans the vote after its 1870 ratification by targeting those identifiers. It was a continuation of the discrimination carried out against them since the birth of the nation.

The Constitution as first drafted treated Native Americans essentially as foreigners—"Indians not taxed" were specifically excluded from the population count when U.S. congressional districts were created, and the Constitution authorized Congress to oversee commerce with Native Americans in the same way it did with other nations. This construction might have made sense if the United States had respected the sovereignty of the Native nations. But the country expanded west, overtaking Native American land and seeking to enforce U.S. law without extending rights to the indigenous people. The arguments made to deny Native Americans citizenship and the vote grew out of a mix of hate and fear. Lawmakers in the 1860s said both that Native Americans were an inferior race and that they had such numbers in certain areas that, if given the right to vote, they could take power. Lawmakers said these things out loud.

Confusion and disagreement over whether Native Americans could be citizens went on for decades, including after the 1887

Dawes Act, which allowed Native Americans to become citizens only if they accepted the land allotment the act gave them or if they chose to live "separate and apart from any tribe" and adopted white people's culture. The 1924 Snyder (or Indian Citizenship) Act finally declared native-born Native Americans as citizens even if they didn't give up their cultural identity, but officials continued to disagree over voting rights. The U.S. Department of the Interior said that Native Americans could vote, but western states, where Native Americans lived in large numbers, said they couldn't.

Native Americans faced a variety of state laws that kept them from the polls. Perhaps the strangest involved the idea of Native Americans being under "guardianship." Some state laws prevent adults from voting if they're under legal guardianship—when their mental or physical capacity requires others to make decisions for them. In 1928, four years after the Indian Citizenship Act, guardianship was at issue in an Arizona Supreme Court case. Two Native American men had been denied the right to register to vote. The Arizona Supreme Court held that those living on reservations had the same relationship with the United States as a ward to his guardian. The opinion stood for twenty years.

The proportion of Native Americans who fought in World War II was larger than that of any other ethnic group, and veterans became an active force in advocating for their people to have the right to vote. A veteran and member of the Navajo tribe is reported to have said, "We went to Hell and back for what? For the people back here in America to tell us we can't vote?"

Veteran, teacher, and master's degree candidate Miguel Trujillo was the named plaintiff in a well-known Native American rights case. In 1948 he was denied the right to vote in New Mexico because that state's constitution barred "Indians not taxed." Trujillo didn't pay property taxes because he lived on a reservation. He did, however, pay plenty of other taxes, including sales and income taxes. The New Mexico federal court sided with Trujillo, ruling that preventing "Indians not taxed" from voting was unconstitutional race discrimination.

After the litigation, Trujillo continued his life as an educator and went on to fight for immigrants and black Southerners suffering under Jim Crow laws. His son, Dr. Michael Trujillo, served as the director of Indian Health Services in the Clinton administration.

Native Americans voted in numbers high enough to be decisive in some elections in western states in the 1950s, including in areas where their right to vote was only recently granted. But universal voting rights and protections continued to be elusive. In 1956, the Utah Supreme Court upheld a law that prohibited Native Americans living on reservations from voting, similar to the way the New Mexico law had. The court found for the state, saying that Native Americans "enjoy the benefits of governmental services" without bearing an equal tax burden "and are not as conversant with nor as interested in government as other citizens." The state legislature repealed the law the next year. Utah was the last state to finally grant Native Americans living on reservations the right to vote.

Scholars and authors who contributed to the book *Native Vote* reviewed the voting-related cases filed on behalf of Native Americans, or cases affecting Native American interests, from the passage of the Voting Rights Act in 1965 through 2006. Of seventy-four cases, seventy resulted in some gains for Native Americans, either because they were decided in their favor or because they were settled on terms favorable to them. The suits involved issues like lack of polling places on reservations, discriminatory election procedures, and methods of electing officials that effectively shut Native Americans out of countywide races.

Daniel McCool, professor emeritus at the University of Utah and a co-author of *Native Vote*, has continued to follow those same types of voting rights cases. He found that the outcomes were favorable to Native Americans' interests around 90 percent of the time in the twenty or so cases filed through late 2019. That tells you two things, McCool said: both that the facts and arguments brought by Native Americans were persuasive, and that a "persistent effort to diminish the rights of Native Americans to vote" continues.

Immigrants

A constant in U.S. history has been a concern that people moving here from other countries will take jobs, bring disease, or change the power structure. Each time a new immigrant group became prominent in the country, changes to citizenship law were debated and often enacted.

At first, gaining citizenship was straightforward in the new nation. The Naturalization Act of 1790 said that "any alien, being a free white person" and "of good moral character" was eligible to become a citizen after living in the United States for two years. But a quarter century after the nation's founding, states began to add citizenship requirements to their constitutions to protect power from shifting to foreign-born voters.

As the nation developed in the mid-1800s, how immigrants were welcomed often depended on the available physical space and land. Predominantly agricultural states like Minnesota, Michigan, Indiana, and Kansas encouraged immigrants and offered them the ability to vote even before they became citizens.

But as the numbers of immigrants in urban areas grew, efforts to limit their rights expanded as well. In 1840, New York passed a voter registration law that affected only New York City, which had a highly concentrated immigrant population.

Nativist tendencies that had been previously tamped down began to gain traction in the mid-1850s. A group called the Know-Nothings gained prominence in the Northeast and parts of the Midwest and South. It originated as a secret society; the name came from critics who would ask assumed members about their activities. The customary reply: "I know nothing." The Know-Nothings supported laws that would impose a twenty-one-year waiting period for naturalization and permanently bar those born outside the United States from becoming officeholders. They also advocated for voter registration systems and literacy tests meant to keep immigrants from being able to vote.

Though the Know-Nothings were never able to change federal voting laws, they shocked what Keyssar called the "political elite" by winning governors' races in multiple states and controlling the legislative branches in six. In 1857, Massachusetts required prospective voters to show that they could read the Constitution and write their names, which the Know-Nothings said would keep "ignorant, imbruted" Irish from the polls. (The Know-Nothings' rhetoric also stemmed from prejudice against Catholicism.)

The second half of the nineteenth century saw growing support for literacy tests and the expanded use of government-issued ballots; both suppressed the vote of those who couldn't read English. Chinese immigrants were among those targeted for their ethnicity. Anti-Chinese sentiment in California led cities and the state to create economic hurdles for Chinese businessmen and to lessen immigration by laborers. A provision in the 1879 California constitution sought to permanently bar natives of China from voting.

Immigrants were dealt a nationwide blow in 1882, when the Chinese Exclusion Act suspended immigration from China and made it illegal for Chinese to become citizens and, thus, have voting rights. These laws stayed in place until the Magnuson Act of 1943, China being a U.S. ally in World War II. Only then were the immigration laws relaxed somewhat—though severe quota restrictions still existed—and some Chinese immigrants given the opportunity to become citizens. It wasn't until the next decade that some Japanese and other Asian immigrants were given the right to become citizens and earn the right to vote by the Immigration and Nationality Act of 1952, though that law also kept a controversial quota system in place.

More than a half century later, Asian American voting rates continue to lag behind those of white and black Americans—65 percent of eligible white Americans voted in 2016, while only half of eligible Asian Americans did. However, candidates are finally beginning to conduct real outreach to the increasingly large voting bloc of Asian Americans. And in the 2018 election, about 40 percent of Asian Americans voted, a 13 percent surge over the 2014 midterms.

If You Can Go to War, You Can Vote: Eighteen-Year-Olds

Gender and class barriers weren't the only voting limitations carried over from Britain. So was the voting age, which remained at twenty-one in the United States until the summer of 1971. A constitutional amendment to lower the voting age was introduced after the draft age was lowered to eighteen in 1942. Those who wanted to honor military service supported it, as did the National Education Association, which argued that the increase in the high school graduation rate meant that eighteen-year-olds were fully prepared for civic life.

The amendment stalled in committee, and the debate over lowering the voting age plodded along until the mass Vietnam protests, when young people were drafted to fight in a war many of them vehemently opposed. Young people questioned if the country would be at war at all if eighteen- to twenty-year-olds could express their political positions at the polls.

Senator Edward Kennedy and two of his colleagues surprised their fellow lawmakers in 1970 by adding language to lower the voting age to a bill extending the Voting Rights Act. The House of Representatives reluctantly supported the bill in order to assure the VRA extension.

In a 5–4 decision later that year, the U.S. Supreme Court ruled that Congress could lower the voting age only for federal elections. The decision presented a logistical nightmare, as many states procedurally couldn't amend their own constitutions in time for the next election. That would mean that eighteen-year-olds could vote in the federal elections on the ballot but not the state ones.

With a swiftness unimaginable today, federal lawmakers moved to amend the U.S. Constitution. The Senate voted in March 1971, 94–0, to make the voting age eighteen. The House quickly approved the amendment as well. By July, enough states had voted in favor, officially making it the Twenty-Sixth Amendment to the Constitution. It was the quickest ratification process in the history of the country.

Roughly four million people turn eighteen every year. As we'll see in Chapter Four, all we need to do now is get them to vote.

CHAPTER TWO

LONG-SUFFERING FOR WOMEN'S SUFFRAGE

Suffragists greet women who have traveled from California to promote women's right to vote, New Jersey, 1915 (Courtesy of the Library of Congress)

California congresswoman Nancy Pelosi took the gavel as Speaker of the House after the 2018 midterm elections, telling the newly sworn-in members of Congress, "I'm particularly proud to be woman Speaker of the House of this Congress, which marks the 100th year of women having the right to vote." For the first time, more than 100 women serve in the 435-member U.S. House of Representatives, and their terms last through the country's August 2020 women's suffrage centennial celebration. Though unquestionably an applause-worthy event, it still means that 100 years after getting the right to vote, the sex that makes up just over half the country holds just under a quarter of the votes in the "People's House."

As conversations about achieving equal representation in government, at work, and at home continue, it's instructive to reflect on how long the fight was for all U.S. women to be able to vote. The women's suffrage movement had been building for three decades before a constitutional amendment to allow women to vote was proposed in 1878, and it would be another forty years before the amendment, named for legendary suffrage advocate Susan B. Anthony, finally became part of the Constitution in 1920.

When pondering a long-delayed victory, it's human nature to focus on the final achievement and not the thousands of actions along the way. We imagine women in formal, floor-grazing skirts gathering for champagne toasts and clinking for joy that they received what is the most basic right of citizenship. And while we should look forward to raising a glass on the 100th anniversary, the celebrations should acknowledge how very difficult the battle for women's suffrage was, and the many complicated but heroic women nearly forgotten. Rarely do we pause to talk about the incarcerations, hunger strikes, and racism involved.

Each new wave of leadership in the movement was built on the work of those who came before—honing their message, taking what worked, and improving upon what didn't. In 1872, Anthony wrote of traveling some 13,000 miles, state to state, and attending 170 meetings the year before to build the movement. With ratification on the horizon in 1920, Carrie Chapman Catt's leather planner detailed states' progress alphabetically, beginning the "A" tab with how many seats were in Alabama's legislature, a note on the length of time specific women had lobbied lawmakers in the state, and the timeline of the attempt to pass ratification there.

The story of women's suffrage is filled with enough magnetic characters, drama, friendship, betrayal, and high fashion to outdo any binge-worthy Netflix period drama. And as with many hard-fought battles in American history, the right thing to do—treating women as equals—was a topic of conversation right from the start.

Neglecting the Women at the Birth of a Nation (Except in New Jersey)

Our first heroine is Abigail Adams, who in 1776 wrote to her husband, the future president John Adams, asking that he consider women's rights as the would-be country prepared to emerge independent from Great Britain. If they didn't, Abigail told him, women would not sit idly by:

> *I long to hear that you have declared an independency—and by the way in the new Code of Laws which I suppose it will be necessary for you to make I desire you would Remember the Ladies, and be more generous and favorable to them than your ancestors. Do not put such unlimited power into the hands of the Husbands. Remember all Men would be tyrants if they could. If particular care and attention is not paid to the Ladies we are determined to foment a Rebellion, and will not hold ourselves bound by any Laws in which we have no voice, or Representation.*

The passage includes some modern spelling tweaks for readability, but John Adams understood the original well. In his reply, he told Abigail her letter was the first "Intimation that another Tribe [women] more numerous and powerfull than all the rest were grown discontented." He then wrote that she was "so saucy" (really). But he continued, "Depend upon it, We know better than to repeal our Masculine systems." He tempered this by saying that men were only powerful in theory; and yet he didn't seriously entertain her request, and in other writings argued against expanding the electorate. One thing we can say about the second president of the United States: he was honest. Also, he was a quote machine, which is great for books but in this case not for democratic progress.

Women were already voting in New Jersey in 1776, a right granted by the state constitution and confirmed in the 1790 election laws. Those laws stated that all "inhabitants" otherwise qualified could vote—interpreted to mean free, property-owning women. The granting of the vote, historians think, was less about conferring rights than about women being a constituency political parties thought they could win. The tides turned in 1807, however, when new wording in the New Jersey Constitution stripped women of the right to vote.

The early 1800s did not see a widespread movement of women demanding the vote. But women could cast a ballot in some circumstances. Unmarried women and widows who owned property were allowed to vote in school-related elections in Kentucky in 1838—education was seen as something that women would particularly care about. This was emblematic of how women were viewed during that time: not as individuals but as mothers and members of a family. The head of the household—the husband or father—was the voice of the house, including at the polls.

Women joined the workforce in much greater numbers as the century progressed, making them not just part of a family but individuals with an income and the challenges that came with working. As the movement to end slavery grew, many women and men who advocated for abolition and for the right of black people to vote came to believe that women should be able to vote as well.

Participating in the movement to abolish slavery also helped many women find their own voices, politically speaking, and they eventually argued for their own rights.

Over Tea, a Movement Is Born

The 1848 Seneca Falls Convention in Upstate New York is widely considered the kickoff of the women's suffrage movement in the United States, though women's rights events, including by African American women fighting to abolish slavery, preceded it.

Seneca Falls is a picturesque waterfront town in the heart of the Finger Lakes region of New York, a drive of four or five hours from Manhattan on today's roads. In the mid-nineteenth century, its townspeople were known to be progressive on social reform issues. Elizabeth Cady Stanton, who would become one of the most prominent leaders of the women's suffrage movement, lived there with her husband, a well-known abolitionist. Thirty-two years old at the time of the convention, Stanton teamed up with her friend the outspoken and popular Quaker minister Lucretia Mott, who was a generation older.

One Sunday in early July following a Quaker meeting, Stanton, Mott, and a few friends gathered at the mansion of their friend abolitionist Jane Hunt, her home likely chosen because she had a two-week-old baby to attend to. At that afternoon tea, the idea for a women's rights convention was born. A notice soon appeared in the *Seneca County Courier* newspaper saying attendees would discuss "the Civil and Political Rights of Women."

Around three hundred attended the convention, on July 19 and 20. Men were asked not to join until the second day. The convention's Declaration of Sentiments proclaimed "that it is the duty of the women of this country to secure to themselves their sacred right to the elective franchise." Additional resolutions stated that women were men's equal; that women should no longer degrade themselves by being satisfied with their position in society; and, my personal favorite: "That the same amount of virtue, delicacy, and refinement of behavior that is required of woman in the social state, should also be required of man, and the same transgressions should be visited with equal severity on both man and woman."

Frederick Douglass, the escaped slave, gifted orator, and leader in the abolitionist movement, attended; he is believed to have been the only black person there. A lifelong women's advocate, he argued that suffrage should be a focus of the movement—something many at the convention disagreed with—and that preventing women from voting was a "repudiation of one-half of the moral and intellectual power of the government of the world."

After Seneca Falls, conventions on women's rights and voting were held regularly in different parts of the country and featured well-known activists. Sojourner Truth, a former slave who became one of the most prominent advocates for human rights, is said to have delivered her famous "Ain't I a Woman" speech at the 1851 Women's Rights Convention in Akron, Ohio. (Historians debate whether she spoke those words, but accounts shortly after describe her speaking eloquently about women's rights.) She also spoke memorably at the 1853 New York City Women's Rights (or Mob) Convention. Hecklers repeatedly interrupted the speakers. Truth got the last word, historically speaking, to those jeering her during her speech. "We'll have our rights; see if we don't . . . You may hiss as much as you like, but it is comin'."

The Civil War brought a stop to the women's conventions. Female activists hoped, as had African Americans, that supporting the war effort might bring an advancement of rights.

Stanton and Anthony held a meeting in New York City in May 1863 and pledged to drive support for federal action to end slavery, eventually gathering hundreds of thousands of signatures. Their formal organization, the Women's Loyal National League, set up offices in Cooper Union in Manhattan.

Breakups, Shake-Ups, and Discrimination

As the movement for women to gain the vote grew and changed, infighting and disagreements led to schisms, notably between those who prioritized securing women the right to vote and those who championed the same right for African Americans.

When the Civil War came to an end, many women were discouraged and angry to see African American rights take priority over women's despite women's efforts during the war. As the Fourteenth Amendment was debated—it passed Congress in 1866 and granted full citizenship rights to men born or naturalized in the United States—women including Anthony and Stanton felt that women's right to vote should have been a larger part of the discussion. They were especially angry when the president of the American Anti-Slavery Society urged fighting one battle at a time, saying it was "the Negro's hour."

Anthony and Stanton were undeniably instrumental in women getting the right to vote, and both fought for the abolition of slavery. But the phrase "never meet your heroes" comes to mind when one learns of some of their actions in the years after passage of the Fourteenth Amendment. Anthony said that the disenfranchisement of women was "as great an anomaly, if not cruelty, as slavery itself." Stanton strongly disagreed with prioritizing men's rights over women's. "Think of Patrick and Sambo and Hans and young Tung [referring to immigrant and minority groups] who do not know the difference between a Monarchy and a Republic . . . Would these gentlemen [statesmen], who, on all sides, are telling us 'to wait until the negro is safe' be willing to stand aside and trust all their interests in hands like these?"

Douglass stayed true to his belief in women's right to vote, though he minced no words in a speech to the American Equal Rights Association in New York in 1869, pointing out the differences between white women's experience and that of black people's. "When women, because they are women, are hunted down through the cities of New York and New Orleans; when they are dragged from their houses and hung upon lamp-posts . . . when their children are not allowed to enter schools; then they will have an urgency to obtain the ballot equal to our own."

A rift opened in the women's suffrage community. Stanton and Anthony created a new organization, focused exclusively on women's right to vote. Prominent suffrage advocate Lucy Stone was among the founders of a rival organization, one that promised to support the Fifteenth Amendment even though it only clarified voting rights for African American men. Stone's organization, the American Woman Suffrage Association, became prominent in the movement, at times more popular than Stanton and Anthony's, because it was considered more inclusive and moderate; it also included male leaders. (Stone is also known for not taking her husband's last name. Married women who kept their last names were called Lucy Stoners.)

Discord among the women's rights factions lasted for more than two decades, with ongoing disagreements over the best way to secure the vote, and both organizations publishing their own newspapers to get out their messages. This split in the movement always gets a lot of attention when the suffrage story is told, but one part of the movement that has historically been underplayed was the way that white women fighting for the vote dismissed and mistreated black women doing the same.

Black women including Truth and Mary Church Terrell, the co-founder of the National Association of Colored Women, participated in the meetings and conventions. Anthony and the prominent journalist and anti-lynching activist Ida B. Wells-Barnett were at times allies but clashed over Anthony's willingness to put up with the racist prejudices of Southerners. In *Remember the Ladies,* writer and editor Angela P. Dodson summed up suffragist groups' uncomfortable relationship:

> *White suffragists did not always welcome this progress by black women and their desire to be working partners in the drive for women's enfranchisement. The national leaders of the movement were willing to sacrifice black support to pacify Southerners and court their support for the ballot.*

Anthony even asked Douglass to sit out the Atlanta women's suffrage convention in 1895, telling Wells-Barnett that she worried that seeing a black man sitting with white women on the platform would be too upsetting for Southerners.

But Douglass never made it to that convention. He would spend his last day, February 20, 1895, at a meeting of the National Council of Women in Washington, DC, where he sat onstage with Anthony. He collapsed that evening, telling his wife about the events of the day. He was seventy-seven.

Despite Anthony and Stanton's indignation about black men getting the vote before women, both were well aware of Douglass's continued contributions to the women's suffrage movement, and mourned and praised him at his funeral, where Anthony spoke and read a statement from Stanton. Carrie Chapman Catt, one of Anthony's younger deputies and a leader of the next generation of the movement, wrote to a fellow suffragist that Anthony and Stanton's praise of Douglass "has completely taken the wind out of our sails . . . [Southerners] were a little suspicious of us all along, but now they know we are abolitionists in disguise."

The tension between suffragists who also fought for African

American rights and those, especially in the Southern states, who believed that black Americans did not deserve equal treatment would never be fully reconciled.

The "Right" to Vote Is Already Right There

In the 1870s, women adopted a legal theory promoted by Victoria Woodhull, the first woman to petition Congress in person. Woodhull argued that because the language of the Fourteenth and Fifteenth Amendments protected citizens' right to vote, women who were U.S. citizens already had the right.

Women in states including Michigan, New Jersey, and New Hampshire attempted to register and vote, and Anthony herself led what the *New York Times* called—under the headline "Minor Topics"—a "little band of nine ladies" to the polls in 1872. A poll worker challenged Anthony but acquiesced when he couldn't think of any legal reason to forbid her to vote. She was soon informed that she would be arrested and, the legend goes, insisted the authorities come and handcuff her. "Well I have been & gone & done it," she wrote to Stanton. Eventually convicted and fined $100, she immediately announced she would never pay. (She didn't.)

In 1878, Senator Aaron Sargent of California, husband of leading suffragist Ellen Clark Sargent, introduced what became known as the Susan B. Anthony Amendment proposing that women be allowed to vote. Nine years later it finally got a vote in the Senate; it was roundly defeated.

As is still true today with regard to certain aspects of election law, states were free to set many of their own rules. Some of the earliest wins for women's suffrage came at the state or even territory level. This is true in the American West, though the reasons for success vary and remain debatable.

The Territory of Wyoming granted women the right to vote in

1869—in part, it's believed, to earn good publicity and encourage women to move there; women kept suffrage rights when Wyoming became a state in 1890.

Utah's territorial legislature gave women the right to vote in 1870, an effort championed by a group of former Mormon men who believed that women would vote to end the practice of a man having multiple wives. Mormon leaders were unconcerned that marriage restrictions would actually happen, but felt it would help other parts of the country think better of the women of Utah.

With the goal of ending plural marriages, the U.S. passed a law in 1887 that applied to the Utah territory. It restricted polygamy and also revoked Utah women's right to vote. But when Utah applied for statehood, Utah women involved in the suffrage movement sought to reverse that ban and ensure the Utah state constitution would include women's right to vote. They worked to have a suffrage organization in most counties, and their efforts paid off: the Utah Constitution gave men and women the right to vote and hold office. In 1896, Utah became the forty-fifth state. Women were able to achieve what they did partly because they weren't fighting the power structure of the state; the Mormon church supported their actions.

Experts are divided on why exactly the western territories moved more quickly to allow women to vote, some believing it was a confluence of the right activists and politicians at the right time, or even that those in power felt that giving women the vote was of little consequence, because women would vote like the men in their house.

University of Pennsylvania political science professor Dawn Langan Teele has argued that the competition between politicians and political parties in the territories and states drove them to grant women the vote—that they saw them as a bloc they'd have a chance to win over. Party affiliations weren't as entrenched as they were in the North and the South, and some politicians believed that women could be recruited as supporters independent of their husbands' choices.

Women vs. Women

Many people, of course, did not want women to have the right to vote, and some of the mobilization against the vote came because people of both sexes feared that women would indeed vote their own minds, especially when it came to Prohibition.

Alcohol severely affected the lives of women whose husbands drank too much. Women married to abusive men or men who were unable to hold jobs—a growing issue in the country that many attributed to alcohol abuse—had no way to protect or defend themselves. It was difficult for women to find work. Married women couldn't control their own finances, and they couldn't prosecute their husbands for abuse or win custody of their children if they did manage to leave. "Wives of drunkards were generally unable to provide for themselves or protect themselves and their children in their own homes. Hence, sobriety became a primary women's rights issue," Dodson wrote. Teele provided a clear example of how the two issues were grouped together, noting that a popular anti-suffrage protest sign read: "We are against Prohibition and Susan B. Anthony. We want our beer, and the men do the voting."

Arguments against granting women the right to vote of course went beyond temperance and gender—race and class were major factors. Southerners were concerned that granting women the right to vote would mean that they had to allow black women to vote, as the more aggressive tactics used against men might be harder for the public to condone if used against women. A 1923 book by suffragists recorded an earlier statement from a Mississippi senator starkly summarizing that position: "We are not afraid to maul a black man over the head if he dares to vote, but we can't treat women, even black women, that way. No, we'll allow no woman suffrage. It may be right, but we won't have it."

The women who fought against their own right to vote remain the most perplexing when viewed from a modern perspective. Many of the women who opposed the suffragists' activism were wealthy, or married to powerful men, or both. (Union army hero General William Tecumseh Sherman's wife, Ellen Ewing Sherman, was one of them.) They enjoyed a social status that protected them from many of the indignities others of their sex suffered, and they worried that if all women could vote, their own position in society would be diminished.

As those women pushed to protect their lifestyles, other outspoken women argued that voting was necessary to lift less fortunate women up. "No one needs all the powers of the fullest citizenship more urgently than the wage-earning woman," labor activist Florence Kelley said in 1898. Kelley later advocated for the women of the National American Woman Suffrage Association (NAWSA) to stop using classist and anti-immigrant language. Changing the rhetoric wasn't just the right thing to do; it was also important to grow the movement.

Parades and Protests

The time had also come to employ more aggressive tactics, ones that would be difficult to ignore. Elizabeth Cady Stanton's daughter Harriot Stanton Blatch returned home to the United States in 1902 after living in Britain and was among those convinced that Americans needed to embrace the tougher methods used by British suffragists. The British fight for suffrage was not the fully rosy approach the mother in *Mary Poppins* projects while prancing through her living room singing "Sister Suffragette." British women were heckling politicians, holding huge disruptive outdoor meetings, and organizing workers to join them in the fight.

One of the most effective changes in the American suffrage movement was the introduction of marches. Women in Iowa and California marched in 1908; the first major parade was in New York City in 1910. In 1912, in a statement likely to sound familiar to the millions of women who marched nationwide in the 2017 Women's Marches, Blatch explained the purpose of the marches in the *New York Tribune* in 1912: "Men and women are moved by seeing marching groups of people and by hearing music far more than by listening to the most careful argument." Demonstrating their numbers and organizational talents would do what logic couldn't. Boosting the visibility of the movement of course had to work hand in hand with the legislative agenda. The jailing and force-feeding some British suffragettes endured also encouraged the movement's American leaders to reintroduce the idea of a federal amendment for women's suffrage. Passing the Susan B. Anthony Amendment was brought back to the forefront of the battle.

Woodrow Wilson, who reportedly told his campaign staff that he was "irreconcilably opposed to woman's suffrage" and that he believed that women active in the movement were "totally abhorrent," became president on March 4, 1913. The women's response to his ascendancy resulted in one of the most memorable images of the movement: twenty-seven-year-old New York University School of Law graduate Inez Milholland, wearing a flowing white cape and a crown with a star, astride a white horse. She rode at the front of a parade of more than five thousand women's suffrage supporters marching up Pennsylvania Avenue in Washington, the route Wilson would take for his inauguration the next day. (Tragically, Milholland died just three years later after collapsing while giving a women's rights speech in Los Angeles.)

The calculated tactics of parade organizers were the subject of a 2019 exhibit at the Library of Congress. "One thing that I find fascinating is the degree to which they took great pride in their banners . . . and what they wore," said exhibit curator Janice Ruth, as she stood in front of rolling black-and-white film of suffrage parades. "They wanted to take what had been more or less traditionally a very masculine streetscape, and they made it their own." Participants often dressed all in white, lending unity to the effort.

Protesters, mostly male and including police officers purportedly there to keep the peace, jeered at the women, blocking their way, and spitting on them. Alice Paul, one of the young suffragist leaders, was delighted with the commotion; she knew a little trouble would get much more attention than a peaceful walk. Rebecca Boggs Roberts, author of a book about the 1913 parade, characterized Paul's attitude: "I don't want safe. I want to march where men march. I want all of the symbolism going through the heart of federal Washington."

Months of planning went into the event. Banners and signs with sayings like "Woman's Cause is Man's They Rise or Fall Together" and the simple slogan of "Votes for Women" are pithy enough for today's Instagram captions. The sign that led the parade was wordy but direct: "We Demand an Amendment to the Constitution of the United States Enfranchising the Women of This Country."

"Paul was amazingly creative in terms of the kinds of publicity stunts that she would come up with," Ruth said. "What would she have done today with the tools that we have available to us now?" Even without the benefit of the social media era, the parade was a success.

With the benefit of a century of hindsight, however, the racism and inattention to inclusivity displayed by the organizers must be noted. Leading up to the parade, Anna Howard Shaw, president of the NAWSA, sent Alice Paul a telegram, saying she'd heard Paul was urging black women not to march, and instructing her to tell the parade marshals to treat black women equally. Some of the white women wanted to have it both ways: they said they supported African American participation in the movement but weren't willing to risk criticism that might hurt their cause. A woman wrote Paul, for instance, saying she was pleased black women were being welcomed, but hoped few would come "in view of the wicked and irrational color prejudice so prevalent."

Come parade day, black women were asked to walk as a group in the back instead of with their own states; journalist Wells-Barnett refused. She marched alongside the white women of her home state of Illinois.

PROS AND CONS OF WOMEN'S SUFFRAGE

Compiled from news sources and advertisements, below are arguments used, on one side, by women's suffrage advocates and, on the other, those fighting against it. Some are funny. Some remain infuriating 100 years later.

For Women's Suffrage	Against Women's Suffrage
"In seven out of twelve suffrage states there is already the Eight-Hour-Day law for minors. Only fourteen of the thirty-six male suffrage states have this law."	"Woman suffrage is 'an essential branch' of the tree of Feminism. To prevent the growth of the tree of Feminism, destroy its 'essential branch.'"
"Women [are] equal in business, professional and industrial life with men."	"Remember that the great majority of women do not want the ballot thrust upon them by the fanatical minority."

For Women's Suffrage	Against Women's Suffrage
"Women have equal educational privileges."	"Remember that woman means suffrage for *every* woman and not only for your own female relatives, friends, and acquaintances."
"There is more democracy in Europe now than in our own country."	"Woman Suffrage goes hand in hand with Feminism, Sex Antagonism and Socialism."
"Why are you paid less than a man?" "Why do your children go into factories?" "Why don't you get a square deal in the courts?" "Because you are a woman and have no vote."	"Woman suffrage brings . . . Jury duty for women."
"There are thousands of children working in sweat-shops . . . Mothers are responsible for the welfare of children. This duty as mothers requires that they should demand VOTES FOR WOMEN."	"Loss of moral influence of good women in public life and increase of political influence by bad women."
"Women bring all voters into the world. Let women vote."	"It would unsex women."
"She's good enough to be your baby's mother and she's good enough to vote with you."	"A Vote for Federal Suffrage Is a Vote for Organized Female Nagging Forever"
"If a woman forges a check, does her father, her husband, her employer, go to jail for a felony? Why is it that the only place in the world where man wants to represent woman is at the ballot box?"	"Women vote would increase ignorant and criminal vote."

Confronting, Picketing, and Persuading the President

Even with the movement's increasing visibility, suffrage remained an uphill climb during the Wilson administration. The Susan B. Anthony Amendment had little to no support in Congress, Southern states remained opposed to black women getting the vote, and the political machines of the North were wary of anything that might upset the status quo.

There was also disagreement within the movement over whether to pursue a state-by-state effort at getting the vote or support the federal amendment. President Wilson, whose political views on the subject had evolved somewhat, spoke out for state referendums over a federal law. Speaking at a women's suffrage convention in 1916, he said, "I have not come to ask you to be patient, because you have been, but I have come to congratulate you that there was a force behind you that will . . . be triumphant, and for which you can afford a little while to wait."

Suffrage leader Shaw responded to the president with a clear demand. "We have waited long enough for the vote, we want it now," she said, reportedly turning to Wilson to address him directly. "And we want it to come in your administration."

The United States entered World War I in 1917, and most suffragists pledged loyalty to the country but continued to demand the right to vote. They picketed the White House daily, carrying signs with messages such as "Mr. President How Long Must Women Wait for Liberty." When counterprotesters assaulted some of the women, it was the women who were threatened with arrest. That year, more than two hundred women were arrested, and some were force-fed after going on hunger strikes. It mirrored what had happened in Britain and brought unwelcome press attention to the Wilson administration.

That July, women flooded the White House with telegrams critical of the arrests. Blatch's July 18 message told Wilson that he should give women the vote instead of jail sentences. The next day, an above-the-fold page-one article in the *New York Times* reported that Wilson was considering doing just that, at least according to the husband of one of the imprisoned women. The New Jersey man met with Wilson and indicated that the president was "deeply shocked over the arrest and imprisonment of the women" and was "anxious to do something to conform to the 'votes for women' sentiment," the *Times* reported.

October 1917 brought the final major suffrage parade. Women marched from Washington Square Park in Lower Manhattan to the Upper East Side in a three-hour procession that included farmers, factory workers, doctors, teachers, and social workers. The next month, New York women won the right to vote when Tammany Hall, the powerful political machine that essentially ran New York City, let the effort move forward. (Tammany Hall boss Timothy Sullivan had supported women's suffrage for decades, even when the movement was staunchly and formally opposed by his colleagues. Sullivan's suffrage support stemmed from seeing the struggles of his mother, who was abused by his alcoholic father. Sullivan died in 1913, so he didn't live to see New York women's suffrage become reality.)

The Constitutional Amendment Comes Down to Tennessee

After New York passed its law, the U.S. House of Representatives again voted on the federal amendment, getting the exact number of votes needed to pass. Nine months later, the Senate voted down the bill.

The leaders of the women's suffrage movement vowed to make

the "no" senators pay, and they did. They campaigned against their reelection and won enough seats to swing the Senate. On June 4, 1919, the Senate approved the amendment. The fate of women's suffrage moved to the states.

An amendment to the Constitution requires approval from three-fourths of the states, so a sprint to get the thumbs-up from at least thirty-six state legislatures began. After thirty-five states agreed to ratify, Tennessee was the last hope to cross the state threshold before the 1920 presidential election.

In the lead-up to the August vote, activists in favor of the amendment and those fighting against it both descended on the wide, marbled lobby of the Hermitage Hotel. Still one of Nashville's finest, the hotel is a short walk from the capitol.

The moment wasn't necessarily marked by dignity or high-minded debate; it was more like a wild rehearsal dinner where a groomsman hopes to sabotage the wedding. Legislators were supplied with seemingly endless alcohol by those who wanted them to vote no—one area of the hotel became known as the "Jack Daniel's Suite"—leaving suffragists to argue their case to lawmakers smelling strongly of whiskey.

As antagonistic as the sides may have been, there were also feminine touches, with suffragists passing out yellow roses and those opposed handing out red ones. The amendment was a nail-biter in the Tennessee House. Harry Burn, a first-time legislator in his early twenties, cast a deciding vote. Burn's constituents wanted him to vote no. He wore a red anti-suffrage rose in his lapel as he prepared to cast his vote. But in his pocket was a letter from his mother telling him to "be a good boy" and vote for women's suffrage. So he did.

A few days later, the amendment was made official, with no fanfare or Rose Garden celebration. August 26, 2020, marks the celebration of 100 years of women's suffrage.

What Women Did with the Vote

Two generations of women had to tune out countless naysayers, challenge authority at the highest level, and be willing to revamp their strategy, decade after decade, to win the vote. They had to use their wits, work the media effectively, and be stereotypically feminine in dress but militant in cause. All for something they should have had since the country's start.

The late, legendary journalist Cokie Roberts put it perfectly when she interrupted NPR reporter Steve Inskeep mid-intro on a story about women's suffrage. Inskeep was describing the passage of the Nineteenth Amendment and its "granting women the right to vote."

"No, no, no, no! No granting!" Roberts said. "We had the right to vote as American citizens. We didn't have to be granted it by some bunch of guys."

For many women of color, the Nineteenth Amendment would make little difference in their lives; they'd have to wait until the 1965 Voting Rights Act to cast a ballot. Even as the amendment was signed, minority women knew what lay ahead. After Texas approved women's right to vote in primary elections in 1918, black women who had gone door-to-door lobbying for suffrage tried to register and were turned away.

Many suffragist pioneers didn't live to see their right acknowledged. Anthony's 1906 *New York Times* obituary said she wanted her last act to be leaving all her money to the suffrage cause. She expressed her intense frustration to the suffragist leader Shaw shortly before she died. "To think I have had more than sixty years of hard struggle for a little liberty, and then to die without it seems so cruel," Anthony said to Shaw.

After the passage of the Nineteenth Amendment, it took American women some time to fully embrace their new right. There was no immediate rush to the polls. There are various theories as to why—women weren't yet in the habit of voting, husbands discouraged wives from doing so, or the majority of women hadn't cultivated an interest in politics. But we eventually grew into our power, and by 1980, the percentage of voting-age women who cast a ballot was higher than the percentage of men who did. And that's been true in every presidential election since.

CHAPTER THREE

VOTING PROBLEMS AND VOTING SOLUTIONS

Climate Strike in Washington, DC, September 20, 2019 (Courtesy of League of Women Voters)

Americans may be divided on countless political issues, but we have a few things we agree on nearly universally. Colorful sunsets demand to be photographed, ice cream is a nice treat on a hot day, and, more seriously, the right to vote is essential to personal freedom.

That last one is something a massive 91 percent of us believe, according to a 2017 Pew study. Since we hold the right to vote so close to our hearts, almost all of us also turn out to vote on Election Day, right? Not even close.

In the 2016 presidential election, 61 percent of eligible Americans voted. Because midterm election turnout typically lags well behind turnout in presidential election years—an abysmal 42 percent in 2014, with some estimates even lower—2018's 53 percent turnout had voting advocates practically shouting in the streets with delight. Still, nearly half the number of people who could have voted in 2018 chose not to or couldn't make it. It was a half-empty glass of democracy.

Why exactly are we so lazy? To start, I don't think we are. We're stretched for time, we're stretched for childcare, and, in some places, limited hours to vote don't accommodate strict work schedules. And even if the majority of us value voting as a right, that doesn't necessarily translate into explaining to younger generations the importance or the mechanics of voting, and that it's a right effective only when put into action. Plus, politicians are locked in ongoing battles over whether or not to make registering and voting more convenient, and which side wins that debate varies from state to state.

Many of our fellow wealthy developed countries have a much higher percentage of their eligible populations voting than we do in the United States. Belgium and Australia, for instance, have turnout rates up to and even over 90 percent. Those two nations' laws require citizens to vote. Still, other countries that don't require voting, but which do have less restrictive voter registration requirements, also have very high turnout.

Jeremy Bird, the national field director for Barack Obama's 2012 reelection campaign, now strategizes for Democratic candidates and progressive organizations. He begins some speeches by asking, "How many people would be here if you had to register for this event 30 days ago? If you had to get a stamp, put it in an envelope and mail it? None of you would be here!" It's a fair point. Having to register well before an election is a factor that keeps people from voting.

Bird's thinking—that registering to vote and voting itself should be made easier—also matches up with the views of a majority of Americans. That same Pew study found that nearly 60 percent of Americans believe that "everything possible should be done to make it easy for every citizen to vote."

What Are the Voting Holdups?

The first thing to understand about the logistics of voting is that the rules are up to each state to set for themselves. Registration requirements, the voting period, and whether people can vote by mail, in person, or absentee are different—sometimes very different—state to state. As historian James Bryce, onetime British ambassador to the United States, wrote, "I cannot attempt to describe the complicated and varying election laws of the different States."

It is clear what Bryce meant. Washington and Oregon vote by mail. My mother lives in suburban Houston. She cast her midterm election ballot during Texas's early voting period, more than a week before I voted on Election Day in 2018, the only day I could vote in New York. (After the 2018 election, New York finally approved early voting.) Some states require IDs, but what counts as proper identification is state-specific; a bank statement works in West Virginia, while a photo ID is needed in Mississippi. That modern-sounding Bryce quote? It's from 1888. We probably should have achieved more uniformity in 130 years.

That lack of uniformity contributes to the differences in voter participation state to state. Maine and Minnesota are state leaders in voter turnout, with numbers regularly over 70 percent. The two states allow voters to register the same day as the election and have minimal voter ID requirements. Minnesota also has a long early voting period. Texas and Tennessee are always near the bottom, turnout-wise, both just above 50 percent in the last presidential election. Those states require registration thirty days before an election. Texans can't register online, and the state requires people seeking to register voters (such as in voter registration drives) to be deputized. Tennessee passed a similar law in 2019 that required training for larger drives and would fine groups who turn in incomplete registration forms. A federal judge called the requirements a "punitive regulatory scheme," and put a hold on it going into effect while a lawsuit brought by voters' rights organizations moves forward. Those groups, including the ACLU of Tennessee, argue the law is designed to stifle minority registration and was created in response to recent drives to register African American voters.

Felons and Voting

How states address voting rights of citizens who have been convicted of a felony also varies wildly, which means states even differ on who is allowed to vote.

Some examples: In Vermont and Maine, felons retain their voting rights even when incarcerated. In states including Illinois, Ohio, and Montana, rights are restored immediately when the person is released from prison. California felons must wait until the end of their parole period (the time left on their sentence if they're released early for good behavior or other reasons). States including Texas and Georgia require felons to complete probation before they can vote. In Iowa, felons are barred for life from voting unless they seek individual exceptions.

Though the decision to take away people's right to vote when they break the law was made on a case-by-case basis even in ancient Greece, the United States is one of only a few developed democracies where some felons don't regain their rights for years after they complete their sentences, and sometimes never do. An estimated 6.1 million American citizens cannot vote because of a prior felony conviction, a 2016 report said. If that group were a state, they'd be about the twentieth most populous.

Many states' broad prohibitions against felons voting arose after the Civil War. Laws that African Americans were considered more likely to break—or laws that were easy to convict African Americans for breaking, whether they'd done it or not—were specifically selected as ones that would result in forfeiting the right to vote. It was another way, often openly acknowledged, to curb African Americans' right to vote.

Florida's 1868 constitution added a ban on felons voting and was designed to keep black Americans from the polls. More than 150 years later, it still disproportionally affected African Americans, with more than one in five prohibited from voting in Florida in 2016. In 2018, Florida citizens—with 64 percent of the vote—approved an amendment of their state constitution to restore felons' voting rights after completion of all terms of their sentence, including parole and probation. Those convicted of certain crimes (including murder) are excluded from the legal changes and will not have their rights restored. Before the vote, ACLU Voting Rights Project director Dale Ho remarked that restoration of felons' voting rights is an issue that has bipartisan appeal; it aligns with Americans' general enthusiasm for second chances.

The victory was widely celebrated by voting rights advocates, who said that it was likely to add more than a million voters to Florida's roll—a big deal in a state whose Electoral College votes can swing a presidential election and whose 2018 governor's race was divided by less than 35,000 votes. It shows how one state's laws can affect the nation; the last three presidential elections in Florida were decided by a margin of about 200,000 votes.

But the celebrations didn't last long. Within a year, the Florida legislature passed a law defining completion of a sentence as paying back all financial obligations resulting from a conviction—not just court fines and fees but also restitution. An individual might be required to pay thousands of dollars, possibly more. Experts believe that thousands—even hundreds of thousands—of would-be voters are likely to be affected. No central state agency tracks fines and fees or their payment, so there's also confusion over who owes what and who is going to enforce the requirement.

Public interest groups sued immediately, saying that the law amounted to an unconstitutional poll tax. In late 2019, a federal judge said that the state could not prevent people from voting if they could not afford to pay the outstanding debts. The federal litigation is still likely to continue well into 2020.

In January 2020, the Florida Supreme Court said the state could consider payments of fines and fees as a requirement of sentence completion. Meanwhile, movement toward restoration of rights for felons is playing out in other states—in 2018, New York's governor began restoring rights to felons on parole; Colorado broadened state laws to allow those on parole to vote; and Nevada changed their law to allow felons to vote as soon as they are released from prison. Kentucky, which placed a lifetime ban on felons voting unless they received individual exceptions, elected a governor in November 2019 who vowed to restore the voting rights of those who had completed their sentences for nonviolent felony convictions.

Some states choosing to restore felons' right to vote while other states continue to bar felons from the polls exemplifies Keyssar's assessment of the country's voting rights journey being one "of inclusion and exclusion . . . and momentum at different places and at different times."

Voter Suppression

Increased exclusion was on the mind of David Remnick, the editor of *The New Yorker* magazine, after the 2018 elections. Voting rights were in the news more than at any time since the civil rights era, he said, due to laws making it harder to vote.

Voter suppression is what it sounds like: something that keeps those who could vote from doing so by making it burdensome or difficult. The 2018 elections brought reports of extensive purging of voter rolls, and of an alleged improper failure by Georgia's secretary of state (who was also in a close race for governor) to finalize thousands of voter registrations. Tribal officials in North Dakota scrambled to handle concerns of Native Americans about a new law that required voter IDs to have a street address; some reservations don't use a formal address system.

Voter suppression in the past was crude and blunt, explained Mimi Marziani, president of the Texas Civil Rights Project. In the days of Jim Crow laws, literacy tests and poll taxes were visible, on-the-books ways to keep minorities and poor people from voting. Even further back, it was simply illegal for those groups to vote. There was "a locked door right in front of you," with no available key, Marziani said.

Current voter suppression is a door that can be opened, but only by following a series of very complicated steps. It doesn't mean voting is impossible, Marziani said, but what's effective about this newer type of suppression is that the harder you make it to vote, the greater a voter's desire has to be to jump over the hurdles. And if that desire hasn't been planted by parents or educators or friends, or one has an inflexible job, or any of the millions of other reasons voting might not be top of mind, those hurdles can be effective.

The barriers can be antiquated registration requirements, closing polling places in areas where there is little transportation, over-aggressive monitoring of voter rolls that force people to re-register or, worse, discover they are no longer eligible when they show up to vote. New voting laws require constant monitoring from voting and civil rights advocates who try to figure out who might be affected and how to help them.

The Story of Voting Rights Act Renewals

Looking at the push and pull over enforcement of the Voting Rights Act gives us a good idea of both how the hurdles come to be and how they are or are not cleared.

The long era of blunt voter suppression was meant to be brought to an end by the Voting Rights Act of 1965. Under what's regularly referred to as the VRA's Section 5, states and counties that had historically disrespected minorities' voting rights had to get approval before they made changes to their voting laws. This enabled the federal government to evaluate and prevent any harmful new laws that would discriminate against minority groups. Lawmakers with a special knack for creating suppressive rules could still make the attempt, but they couldn't put their plans in place without some oversight.

"The right to vote is the crown jewel of American liberties, and we will not see its luster diminished," Ronald Reagan said in 1982, when he signed the Voting Rights Act renewal. It's a quote regularly trotted out to show the country's dedication to protecting the sanctity of voting. But like so much else, the inspiring quote doesn't exactly fit with the often dicey backstory of the enforcement and renewal of the VRA.

Ari Berman's *Give Us the Ballot* uses congressional testimony and governmental memos evaluating whether to renew the VRA to show that attempts at suppression never really stopped. In 1969, President Nixon and Southern legislators had plans to dilute the

VRA before its renewal, but the Senate pushed through a "compromise" that officially outlawed literacy tests nationwide and kept Section 5 intact. Nixon surprised many by allowing the compromise to become law, but only after his chief speechwriter told him that vetoing the bill would be a political mistake. "We should bear very clearly in mind that whatever their views on other black claims—housing, jobs, schools, etc.—the right to vote is one thing that practically all Americans agree on," Ray Price wrote.

The VRA paved the way for Andrew Young, a former aide to Martin Luther King Jr., to become Georgia's first black member of the U.S. House of Representatives in 100 years. It took a lot of shoe-leather campaigning, and enforcement of the Voting Rights Act. His 1972 victory came after the federal government forced a redrawing of the boundaries of Young's congressional district to address concerns of racial discrimination, and after Young sued under the VRA to challenge polling location changes disproportionately affecting black voters.

Just as immigrant groups suffered discrimination in trying to attain citizenship and get the vote, many were also subjected to voter suppression. In the mid-1970s, testimony from Modesto Rodriguez, a Mexican American, led Texas to be added to the list of areas that required pre-clearance for voting law changes and an expansion of the VRA to cover language minorities. Modesto testified that Frio County officials and local employers were economically intimidating Mexican Americans by refusing to give them business loans if they were politically active, and that election materials available only in English were effectively literacy tests for many citizens from that part of Texas. Rodriguez was savagely beaten by officers from a Texas state agency shortly after federal lawyers left his home county after visiting to investigate the discrimination.

While Jimmy Carter's administration brought support for the

VRA, the tide almost turned during the Reagan years, in part due to a talented twenty-six-year-old government lawyer named John Roberts. In the lead-up to a 1982 vote on whether to reauthorize the VRA, Roberts and an assistant attorney general argued that it shouldn't matter if the results of a voting law were discriminatory, as long as the intent of the law wasn't to be discriminatory. Their point was that those objecting to a law would have to show both that the law was harmful and that the people who wrote it meant to cause harm.

That argument played out as a group of congressmen, including Henry Hyde of Illinois, traveled to Alabama to hear testimony about voting conditions in the South. Among those testifying was fifty-year-old teacher and activist Maggie Bozeman, from Aliceville, Alabama, in Pickens County. Bozeman pointed out that the polls in the 1980 election were open only from 9:00 to 4:00, when working-class people couldn't make it; that election officials visited the homes of every black person who had requested an absentee ballot, to make sure they were out of town; and that she'd been photographed by a deputy sheriff's officer when helping illiterate people at the polls. "Just being a voter in Pickens County is a wearying experience," she told lawmakers.

Hyde had introduced legislation to eliminate Section 5's preclearance requirement, and was shocked by Bozeman's testimony and that of others with similar experiences. He changed his mind on Section 5. "I was wrong and now I want to be right," he said. Congress eventually settled on language that would look to the results, not the intent, of potentially discriminatory laws. Denying the restriction favored by the young John Roberts, Reagan signed the renewal in June 1982. It was on the occasion of the signing that he called the right to vote the "crown jewel of American liberties"—"As I've said before," he underscored.

Fast-forward thirty-one years, almost to the day. Roberts, no longer a young lawyer and instead the chief justice of the United States Supreme Court, got another chance to reshape and restrict the VRA.

Roberts's Rules

In 2013, Chief Justice Roberts wrote the majority opinion in *Shelby County v. Holder,* in which the Supreme Court "effectively struck down the heart of the Voting Rights Act of 1965 by a 5-to-4 vote, freeing nine states, mostly in the South, to change their election laws without advance federal approval," reported the *New York Times*'s Adam Liptak. The chief justice emphasized his belief that some of the restrictions in the act had outlived their time. "Our country has changed," Roberts wrote. "While any racial discrimination in voting is too much, Congress must ensure that the legislation it passes to remedy that problem speaks to current conditions."

While discriminatory voting laws are still illegal, states with a history of discrimination can now enact laws without oversight. Citizens affected by new laws can still challenge them, but cases have to work their way through the court system. The nickname for *Shelby County* should be the "Lawyers' Full Employment Act," Native American rights advocate and scholar Daniel McCool said. "Now you just have to sue them over and over again."

You didn't have to wait long to see the impact of the decision. In 2012, Texas had wanted to enact a voter ID law that was opposed by the Texas NAACP and the Mexican American Legislative Caucus, but, before *Shelby*, a federal judge found that the state could not show that the bill would not discriminate against African American and Latino voters. Hours after the *Shelby* decision, the Texas attorney general announced that the law would take effect immediately. A few months after the 2013 *Shelby* decision, North Carolina also instituted a photo ID bill, shortened its early voting periods, eliminated early voting on Sunday, and ended same-day registration and preregistration for high schoolers.

The Texas and North Carolina laws became the subject of lawsuits. A federal appeals court in Texas ruled that the ID law violated Section 2 of the VRA. Texas later passed a new voter ID law; that newer law survived litigation against it and is still valid. In North Carolina, a legal saga played out over three years. Eventually, the U.S. Court of Appeals for the Fourth Circuit in 2016 struck down the new legislation, in a ruling cited in numerous voting rights lawsuits since. Lawmakers had "requested data on the use, by race, of a number of voting practices," the appeals court ruled, adding that North Carolina had targeted "African Americans with almost surgical precision." The U.S. Supreme Court declined to hear the case, so the Fourth Circuit opinion was the final word. (North Carolina also later passed a modified voter ID law allowing for more forms of ID that was scheduled to go into effect in 2020, though in December 2019 a North Carolina federal judge, citing the state's "sordid history" of voter suppression, said it will not go into effect until litigation over it is resolved.)

Voter suppression is something most of us think about only around election time, but fighting it is something that occupies lawyers and voting advocates year-round. Ho, of the ACLU, was part of the team that fought the North Carolina changes. He has overseen lawsuits throughout the country, including the challenge to a citizenship question on the census that went to the Supreme Court in 2019.

Ho notes that efforts to suppress the vote revved up a few years into the Obama administration, even before *Shelby County.* He points to a report by New York University School of Law's Brennan Center for Justice that found that from early 2011 through

October 2012, nineteen states passed restrictions that could make it more difficult to vote, including voter ID laws, limiting early voting periods, and ending same-day registration options. Some of these laws were overturned by the courts, some were pushed back by voters, and some were upheld by the courts. But their number shows that adding new barriers to voting was top of mind for some state legislators.

CONFLICT AND CONFUSION OVER VOTER ID

Broadly speaking, Republican politicians like the idea of requiring identification to vote, as a way to prevent voter fraud; and Democratic politicians are against it because they believe it suppresses the vote. The majority of the American public favors requiring IDs to vote. To many, the requirement does not seem stringent. It seems like no big deal. Most of us have driver's licenses or a government-issued ID of some sort. We need it to fly, pick up packages at FedEx, and check in at security in many urban office buildings. As of late 2019, thirty-five states require ID on Election Day, with seventeen of those requiring a photo ID. The other states require that you identify yourself, usually by signing in the voter roll next to where you signed when you registered to vote.

Those who take issue with voter ID laws largely focus on the ones requiring a photo ID. The laws sometimes seem specifically targeted to make it more difficult for particular groups to vote, such as young people, when college IDs are prohibited. The young, the elderly, the poor, and minority groups are the most likely to lack proper ID.

Still, studies have shown that those impacted represent a relatively small portion of the population, an even smaller portion of whom are likely to vote. An analysis of Texas voters—Texas constantly being embroiled in voter ID litigation—found that 4.5 percent of registered voters in the state lack proper identification, though only 1.5 percent showed up to try to vote in 2012. (Today, 4.5 percent of registered voters would be more than seven hundred thousand Texans.)

Analysis from the data crunchers at FiveThirtyEight came to the conclusion in 2018 that voter ID laws likely do not have a large impact on turnout. But "even if voter ID laws haven't swung election outcomes, they can deny thousands of people their right to vote—denials that fall disproportionately on black and Latino citizens," wrote Dan Hopkins, a professor of political science at the University of Pennsylvania.

In 2019, a study analyzing nearly a decade's worth of election data found that voter ID laws didn't affect registration or turnout in a measurable way, but neither did ID laws serve their presumed purpose of preventing voter fraud, as there was very little voter fraud to stop. The overall point of the study was that the ID laws don't serve much purpose, which leaves one to wonder why states are so intent on passing them.

Monitoring Voter Rolls

The League of Women Voters, which celebrated its 100th anniversary in February 2020, is one of the most prominent organizations fighting voter suppression, both in the courts and via grassroots outreach through a national chapter and more than 750 state and local ones. It is a nonpartisan group that works to educate voters and encourage participation in government. The league has a distinguished history and enjoys high standing in a range of communities. Because of its reach, it is able to stay on top of proposed voting and election law changes. It endeavors to voice any concerns before a law passes, and to find remedies that will head off future suppression. If necessary, the league sues to protect voters' rights.

One issue that the league is watching closely is the "purging" of voter rolls. States and municipalities want to have clean voter rolls, meaning that everyone listed as eligible to vote should actually be there: people need to be removed from the rolls if they die or move, and laws require states to stay up-to-date. Some states have gotten aggressive about culling their rolls, and the methods can have a disproportionate effect on minority voters.

A July 2018 study by NYU's Brennan Center found a 33 percent increase in the number of voters purged from state rolls between 2006–2008 and 2014–2016. An updated evaluation, in 2019, found that "between 2016 and 2018, counties with a history of voter discrimination have continued purging people from the rolls at much higher rates than other counties." In other words, counties and states that previously required approval for changes in voting laws are the ones doing most of the purging.

States are more emboldened to cull their lists following the 2018 Supreme Court decision *Husted v. A. Philip Randolph Institute.* The *New York Times* described it this way: states can "kick people off the rolls if they skip a few elections and fail to respond to a notice from election officials."

If you don't vote in two rounds of federal elections (you didn't vote in the 2016 presidential race or the 2018 midterms, for example) and fail to return the postcard the state sends you, the next time you show up to vote, you might discover you aren't registered.

Not all states are this aggressive, but it's a good reminder to stay diligent.

VOTING TIP: The best way to avoid being incorrectly taken off the rolls is to keep your information up-to-date and not skip elections. It's also always best to confirm your registration before each election, prior to when your state's registration period ends.

In a similar scenario, some voters found that their names never made it to the rolls. In the lead-up to the 2018 Georgia governor race, 53,000 voter registrations were reportedly on hold because they were flagged in a controversial "exact match" review overseen by secretary of state and gubernatorial candidate Brian Kemp. If the name that a voter had used to register didn't perfectly match some other state record, like a driver's license, his or her registration might be called into question. Of people flagged, 70 percent or more were estimated to be African American, a group Democratic challenger Stacey Abrams specifically focused on in registration drives. Kemp came out ahead by 55,000 votes; he denied any wrongdoing and said that he had not been involved with the review.

Last summer, Abrams launched Fair Fight 2020 to fight voter suppression in battleground states, focusing on training local teams to respond to issues Democratic voters face when trying to register or vote, including monitoring the purging of voter rolls via methods like "exact match."

"Voter suppression is tricky, because the pieces that are put in place to lead to voter suppression happen in between elections," said Celina Stewart, senior director of advocacy and litigation for

VOTING TIP: If you've registered but show up to find you aren't on the roll, fill out a provisional ballot and ask poll workers what you need to do to follow up and correct any error. Your vote could still count.

the league. The impact may not be obvious, she said, until the registration process or Election Day itself, when voters who were unaware of changes to their polling location or shortened registration periods discover them too late. Imagine waiting an hour in line just to get to the front, learning you're in the wrong place, and going to the right one only to discover a four-hour line.

Limitations on early voting and reductions in the number of polling sites are sometimes attributed to budget cuts. But laws requiring stricter identification to vote or restricting registration options are almost always suggested as a remedy for one malady: voter fraud.

The Myth of Rampant Voter Fraud

Voter ID laws, the purging of voter rolls, and limits on making registration more accessible are often done in the name of preventing voter fraud. Voter fraud—people pretending they are someone else when they vote or casting a ballot when they know they aren't allowed to—is obviously wrong and a gross misuse of the system. It's also extremely risky behavior: punishment can be thousands of dollars of fines and even jail time, as well as deportation for noncitizens.

But voter fraud of that sort rarely happens. There are few cases of people being prosecuted for voter fraud. Often, people who have attempted to vote when they shouldn't have made a mistake—

for instance, thinking that they could vote because they had a green card. Even in cases where people who shouldn't vote make it through the checks and vote, the total number represents a drop of water in the ocean of ballots cast.

Giving voter fraud much oxygen feels like contributing to the problem. But here are some stats from litigation involving former Kansas secretary of state Kris Kobach, who maintains that voter fraud is an active threat to our elections. He and Kansas were sued in federal court over a law that required Kansans to show proof of citizenship to register to vote, and one of the issues the court reviewed was whether large numbers of non-citizens were in fact registering. Kobach argued the case himself. Dale Ho led the team for the ACLU.

According to ProPublica reporter Jessica Huseman, who closely covered the trial, the best Kobach could do was show that "over a 20-year period, fewer than 40 non-citizens had attempted to register in one Kansas county that had 130,000 voters." Most of those forty were found to have made a mistake rather than committed intentional fraud, and only five actually cast a ballot.

Texas also made a mountain of trouble for itself while trying to fight the molehill of voter fraud. In an effort to remove what the state said were non-citizens from the voter rolls, the acting Texas secretary of state in 2019 notified county officials that the state had flagged 100,000 people on the various rolls who, when applying for a driver's license or ID card, had said they weren't citizens. The state also referred those people for possible criminal prosecution.

Within days, Texas officials acknowledged that thousands of people on the list had become naturalized citizens and thus were eligible to vote. The state had simply failed to check for this possibility. In addition, about a quarter of the people on the list were there because of confusion between state offices. The League of Women Voters was among several groups that sued the state to stop the purge, and a federal judge put a hold on removing people from the rolls while officials tried to sort things out. In the month after the rollout, forty-five non-citizens had taken themselves off the voter rolls; ten were shown to have voting histories, the *Texas Tribune* reported.

What to Do to Make Sure You Can Vote

Understanding that voter suppression does occur is important, said Jeanette Senecal, senior director of mission impact for the League of Women Voters. But she worries that too much focus on suppression might actually suppress the vote more. Her message to potential voters: if you confirm you're properly registered, know available voting methods and locations, and give yourself time to get to the polls and wait in line, if necessary, then you can vote.

Before Election Day, follow the registration and voting procedures for your state. There are plenty of websites to help you. Each state has a secretary of state website that includes voting regulations, but free private services like TurboVote will go the extra mile and text or email you. Many other get-out-the-vote organizations offer help, too. This is a rare instance when you want to opt in for email or text reminders—they'll keep you up to speed about deadlines and election days.

The League of Women Voters also has an information site, VOTE411, which has been running for more than a decade. In the 2018 midterm cycle, more than five million people visited, the site's biggest year ever. "We know that the top three things that voters actually ask about are: Am I registered? Where's my polling place? And what's on my ballot?" Senecal said. VOTE411 has that information and much more, including registration dates, upcoming election dates, and how long early voting periods are.

People also need to know what ID they need, what kind of voting machines they might see, and what the absentee ballot requirements are. "So we'll push those forward as alerts as well, just so we're helping people get to information they may not realize that they also need to know," she said.

VOTING TIP: We don't always know what we need, so make it easier and let the experts help you. Sign up for those alerts!

VOTING VOCABULARY: RANKED-CHOICE VOTING

In elections with more than two candidates, voters often have to gamble.

If they prefer one candidate and don't have fundamental concerns about the others, the choice is simple. But if they would prefer nearly anyone else but one of the candidates, their best option is to guess which of the candidates is most likely to beat the one they most dislike.

Ranked-choice voting addresses that problem, and in June 2018 Maine became the first state to use it in a statewide election. In the same election, voters were also asked to choose if the state should continue to use it in the future. "This is a little bit like Luke Skywalker blowing up the Death Star. You get one pass," said Matt Dunlap, Maine's secretary of state and the man in charge of making sure the experiment was successful.

Here's how it works: If there are four candidates (let's

call them Aubrie, Bryan, Carson, and Debbie), voters mark which one is their first choice, second choice, and so on. When the votes are counted, if one of the four receives a majority of the votes, he or she wins. But if no candidate gets more than 50 percent of the vote, the fourth-place finisher is eliminated. If Debbie finished fourth, she's out of the race, but everyone who voted for her still has a voice, because their second-choice votes will be counted. Those votes are distributed to the appropriate candidate, and if that count is enough to give Aubrie, Bryan, or Carson a majority, that person wins. If not, the process continues.

Proponents of ranked-choice voting say it gives voters more flexibility, saves money on runoff elections, and can discourage negative campaigning. The idea is that candidates will want to be voters' second choice if they can't be first, and thus won't go too negative about a voter's first choice. (This theory seems very candidate-dependent, but is a nice idea.) Those against ranked choice say it's too confusing for voters and that it still can fail to produce a majority winner when ballots get "exhausted" by voters not having ranked enough candidates.

After the first try with ranked choice, the voters in Maine elected to keep it.

While ranked voting is new to statewide races, it's not a new concept. Cambridge, Massachusetts, has used it since the early 1940s, and versions of it date to nineteenth-century England. Many U.S. cities, including San Francisco, use it for local elections, with more regularly adopting or exploring the concept. New York City voters in 2019 adopted ranked choice for some city elections.

Kansas and Hawaii were expected to use ranked-choice voting in 2020 Democratic primaries, with a few other states considering it as well. Maine is expected to skip ranked-choice voting for the presidential primary but will use it for the 2020 presidential election.

Making Voting Easier: AVR and SDR

Voting advocates point to two things that can best minimize registration-related problems: automatic voter registration (AVR) and same-day registration (SDR).

Automatic voter registration occurs when eligible state citizens interact with a government agency—usually the Department of Motor Vehicles—and the relevant information they provide is transferred electronically to election officials. In some AVR states, citizens opt in to be automatically registered, but in most states, people must opt out. If voters update their personal information, such as when they move, their voting info is also updated.

AVR is a relatively new phenomenon. In 2016, Oregon was the first state to give it a try. Since then, more than a dozen states have adopted it, either via state citizens voting for it or by state legislatures passing laws in favor of it. The benefits of AVR go beyond convenience for the voters; there's plenty in it for the states as well.

"It's a cost savings. It's efficiency. It results in more accurate rolls," said Paul Gronke, a Reed College political science professor in Portland and an expert on electoral behavior and voting laws. If you believe there's a danger of ineligible people somehow getting on the rolls, this feels like a way to prevent that, he said.

The 2016 presidential election was the first after Oregon instituted AVR, and its benefits were notable. Gronke and his colleagues studied the early implementation of AVR in that state, and found that nearly 100,000 citizens newly registered through AVR voted in the 2016 presidential election, and that people who registered through AVR made up nearly 9 percent of Oregon's voter rolls. It also seems to have amplified young voters' impact—Oregon citizens aged eighteen to twenty-nine make up 20 percent of the state but were 37 percent of AVR voters in 2016. Though turnout can't all be attributed to AVR, it certainly didn't hurt: 70 percent of Oregon's voting-age population voted, a record.

That younger citizens are taking advantage of AVR makes sense, as they are most likely to interact with driver's license offices for the first time. But AVR registrants were also more likely to live in suburban areas, low-income areas, low-education areas, and racially diverse areas, the Oregon study found. So at least in Oregon, AVR reached voters who were likely to fall across the political spectrum. "It moves partisan competition to the right place," Gronke said, meaning it lessens partisan fighting over registration hurdles and focuses the competition on the actual race.

In 2018, automatic voter registration was on the ballot in Michigan and Nevada, both considered swing states. In Michigan, about 66 percent of voters approved the method, and in Nevada, 60 percent of the voters approved. When AVR went into effect in fall 2019, registrations immediately went up.

The Brennan Center at New York University released a first-of-its-kind study in April 2019 looking at the effect of AVR on voter registration rates. The study was conducted over a period when registrations are generally quiet; in this way, it could measure how much AVR—rather than increased registration drives near an election, for instance—affected rates.

Evaluating seven states and Washington, DC, the study found increased rates in all. There was a 16 percent increase in Colorado, a nearly 27 percent increase in California, and a 34 percent increase in Alaska. Georgia saw an astonishingly high 94 percent increase in the time period observed. The discrepancy between Colorado's and Georgia's numbers, said Kevin Morris, a qualitative analyst and one of the study's authors, can be explained in the following way: Colorado already has registration-friendly procedures and outreach, so there are fewer unregistered people; Georgia has many strides to make in registration and also aggressively purges voter rolls, requiring more people to re-register. "AVR can do a lot in a state like that," Morris said.

Registration doesn't guarantee voting, and in Colorado, Oregon, and Vermont, voters who sought out registration, rather than being registered through AVR, were more likely to actually vote, a FiveThirtyEight study found. Rhode Island saw non-AVR and AVR registrants vote at similar rates, and AVR registrants in Washington, DC, were actually more likely to vote. It means, of course, efforts must be made to get new registrants to vote—but a person certainly can't vote if they're not registered.

As with most voting issues, AVR is not without its skeptics, who are concerned that failures by government agencies could result in non-citizens being registered. California came under heavy criticism during early implementation of its AVR program when the Department of Motor Vehicles admitted, in 2018, that it may have registered 1,500 non-citizens and improperly classified the preferred political party and vote-by-mail preferences of about 23,000 citizens. (The DMV said the mistake was theirs; the non-citizens had not sought to be registered, and the error did not result in non-citizens actually voting.) In April 2019, a *Los Angeles Times* editorial said while kickoff of the program was "a disaster," it was fulfilling its promise a year in, with one million new voters registered.

Fearful of similar problems due to its outdated technology, West Virginia has delayed its own implementation of AVR. Officials say the system cannot handle the AVR process, and that county clerks would have to manually check any new voter registration.

States (and state budgets) must address outdated technology. Nevertheless, this issue can't detract from the fact that where AVR is instituted, it works. As Brennan Center official Myrna Pérez said when the 2019 study was released: "We should be making it as easy as possible for eligible citizens to vote, and that begins with getting registered. Our current voter registration systems are onerous and outdated . . . and our election systems are overdue for an upgrade. As states continue to enact restrictive voting laws, AVR is a needed change."

Another change to consider? Allow same-day registration. It's exactly what it sounds like: citizens can register and vote on the same day. More than a quarter of states already do this, with Maryland and Michigan voters electing to add themselves to the list in 2018. Maine, Minnesota, and Wisconsin have had same-day registration since the 1970s. Minnesota, in fact, has many things it can brag about voting-wise; in addition to regularly leading the nation in turnout, it holds the all-time state turnout record, at 78 percent (2004). In that election, 20 percent of people who voted in Minnesota registered the same day. Beginning in 2020, Californians will be able to register on Election Day at local polling locations.

For people who have been hesitant to participate in the electoral process, or find themselves moved at the last minute by a candidate, it's helpful. "It's easy for voters to understand [having] to register and to vote the same day," the league's Stewart said. "Those are the things that we have to think about when we're thinking about remedies" for voter suppression.

A point that's stuck with me throughout my research on increasing turnout was one made by Jeremy Bird, the former Obama staffer who called expanding AVR "an obsession." Enacting things like AVR nationwide could directly affect the way campaigns are run. "We spend tens of millions of dollars, probably hundreds of millions, registering voters," said Bird. "We could spend that money educating people and getting them out to vote."

A Push for Vote-by-Mail

Historically, states handle voting by mail, also called vote-at-home, differently. In addition to Washington and Oregon, as of early 2020, Colorado, Hawaii, and Utah mail ballots to all eligible voters, and California increasingly uses vote-by-mail. Following the United States' outbreak of the novel coronavirus, voting experts called for all states to adopt statewide vote-by-mail to avoid disrupting the 2020 election.

All states already allow vote-by-mail for absentee voting, though about a third require an "excuse," like being out of town on election day. Moving to an all vote-by-mail system may challenge states unused to processing a lot of mail-in ballots. "Every state could do this," Washington secretary of state Kim Wyman said in March 2020. But, she cautioned, "It takes time to ramp up."

Wyman, a Republican, said that voting by mail has bipartisan support in her state, but that the base of her party generally does not favor it. Even those who propose expanding vote-by-mail say success requires educating voters and instituting plans to notify those who return ballots with defects. Officials must also consider those who have struggled to use vote-by-mail, such as voters requiring translation services or who lack convenient post office access.

How nationwide vote-by-mail would affect turnout is uncertain, but it could be good news. In 2016, voter participation in vote-at-home states was 10 percentage points higher, on average, than other states, a report by the left-leaning Center for American Progress said. Those rates aren't all attributable to vote-by-mail, but studies generally show increased turnout where vote-by-mail is used. The brightest statistic may be out of Colorado after its vote-by-mail implementation, which also allows election day voting and ballot drop-off. Voters ages eighteen to twenty-four outperformed their 2014 turnout expectation by more than 10 percent.

PART TWO

How to Get People TO VOTE

CHAPTER FOUR

TRANSFORMING NON-VOTERS INTO VOTERS

Yara Shahidi at the We Vote Next Summit in Los Angeles, California, September 29, 2018 (Paul Archuleta/Getty Images)

On a predictably pretty Los Angeles morning, *Grey's Anatomy* and *Scandal* creator Shonda Rhimes told an audience of ten thousand young women they were a "tribe" who needed to vote if they want their "voice to be represented." That afternoon, former

Barack Obama staffers previewed their *Pod Save America* HBO special at a landmark art deco theater in nearby Glendale. It included a video promoting door-to-door canvassing to get out the vote. The next morning, teenage TV star Yara Shahidi gathered more than a hundred young activists at an open-air office park not far from Los Angeles International Airport to brainstorm. The question was how they could get their peers to go to the polls. A few weeks later, at the storied Cooper Union in New York City, where suffragists had once lobbied for women's right to vote, phone bank and canvassing volunteers were rewarded with a pep talk by *Avengers* franchise star Mark Ruffalo as they headed into the final weekend before the November 2018 midterms.

The inescapable message was that voting is not just important; it's also empowering and something to do with your friends. The second continually reinforced truth? Convincing a person to vote for the first time requires demystifying the process, from how to register, to learning about candidates, to finding where to vote.

All of these events involved organizations created after the 2016 presidential election to encourage Americans who historically haven't voted to get in the game. Their collective efforts with fellow get-out-the-vote organizations, along with the general political climate, made a difference. In 2016, turnout for voters between the ages of eighteen and twenty-nine was about 46 percent, within a few points of where it's hovered, save a few bigger dips, since 1974. That age group's turnout is even worse in nonpresidential years: in the 2014 midterms, only about 20 percent of potential voters cast a ballot. More than 35 percent came out in 2018, a significant jump.

Voting advocates hope that improving on the get-out-the-vote methods used in 2018 will lead to record turnout in 2020. In our very virtual world, making a personal connection is still the most effective way to get someone to vote. The idea is this: harness the modern tools of social media to employ the age-old tactic of peer pressure to get friends, both online and IRL, to complete the analog task that is voting.

Capturing the Youth Vote Is Getting Old

National movements to increase youth turnout started decades ago, a fact not lost on politicians, voting advocates, and celebrities who've dedicated a lot of effort to the task. Members of Generation X are likely to recall the genesis of Rock the Vote, founded in 1990 to encourage youth participation in politics and still going strong. Its first partnership was with MTV, whose town halls for young people became a thing of legend when candidate Bill Clinton in 1992 said he'd "inhale" marijuana if he had it to do over again, and then in 1994, when an audience member asked President Clinton if he wore boxers or briefs.

Shortly before performing at a "Party at the Polls" voting event in Nashville in late 2018, rock star and get-out-the-vote veteran Sheryl Crow talked youth turnout as her preteen son tinkered with instruments nearby. Crow began working with Rock the Vote shortly after its start and in 2008 toured with the Beastie Boys on a Rock the Vote tour. She's among the activists and researchers who hope this youth movement is different; they expect—and very much hope—young people will be galvanized by issues they feel affect them directly, including gun safety and climate change. She'd felt there was momentum from youth voters in the past, she said, but sees a new urgency. "This feels as though we're at a turning point."

Disliking the candidates and issues was the top reason young registered voters gave for staying home in 2016. According to a report by CIRCLE, an organization that analyzes youth voting, political "parties and campaigns are having trouble connecting with and motivating youth across the board." But that was before the February 2018 mass shooting at Marjory Stoneman Douglas High School in Parkland, Florida, in which a former Stoneman Douglas student killed seventeen people, including fourteen students. The hugely visible and emotional demand for gun reform by a group of the school's surviving students became a movement. Leading up to the 2018 midterms, 77 percent of young people who identified themselves as likely voters said gun control would be an important issue in determining their vote. Overall, 64 percent of young people favored stricter gun control laws in the United States, a 15-point increase from a similar poll conducted after the devastating 2012 shooting at an elementary school in Newtown, Connecticut.

Helping potential voters connect policy change to voting still requires pointed effort from voting organizations. Many young people don't learn enough about voting before they become eligible, and it's tough for them to feel that their vote counts, given that politicians don't tend to pay them much genuine attention.

Paul Gronke, the Oregon-based professor, is straightforward with his students about politicians being nonresponsive to constituents who don't go to the polls. You have problems getting your student loan concerns addressed, he explains, yet his social security and tax deductions are always a priority for politicians. Guess why? he asks. "Because I vote all the time and you don't."

Those who work to register high schoolers often employ a similar tactic, using age-specific statistics to demonstrate the political influence young voters could harness. Hannah Mixdorf, who helps lead youth voting organization Inspire2Vote, makes sure students know that Americans eighteen to twenty-nine are a big chunk of eligible voters but vote at the lowest rate. She encourages them to see the potential. "Imagine the power you'd have if this entire generation of voters mobilized and cast a ballot."

Speaking Their Own Language

The only way to get young people to care about voting is to earn that enthusiasm—to show why it's relevant to their lives, to make sure they understand the mechanics of how to do it, and to use language they relate to in explaining it.

Nowhere was that sentiment more present than at Yara Shahidi's We Vote Next Summit, held at the TOMS shoes headquarters in Los Angeles. Shahidi is an actress on ABC's hit show *Black-ish* and stars in and produces its spin-off *Grown-ish*, has more than four million Instagram followers, has appeared on the cover of more than a dozen glossy magazines, and is widely featured in their online counterparts. She is a student at Harvard, and Oprah has predicted she'll be president one day. When Barbie created a Yara doll in 2019, in celebration of its sixtieth anniversary, the doll was wearing a t-shirt that read "VOTE" in multicolored letters.

In early 2018 Shahidi appeared on *The Late Show with Stephen Colbert* and told him that her upcoming eighteenth birthday celebration would be a "voting party." She'd get her friends to register to vote. Her organization, Eighteen × 18 (transitioning to the name We Vote Next for 2020), is that idea on steroids. She's created a community focused on educating young people about the voting process, and helping them see it and political action as a regular part of their lives—in general making it "the slightest bit easier for our generation to vote," she said at the 2018 summit. Their Instagram feed provides voter registration tips and information nearly daily, including a heads-up about city, state, and special elections.

But she pushes the big-picture importance of voting, too, working to convince her peers that failing to raise their voices through voting means giving up control of their generation's destiny. Gathering teen thought leaders together is one way she does it, knowing they'll simultaneously spread the message to their own online and real-life communities.

We Vote Next attendees were welcomed by five-foot-tall yellow-and-white signs highlighting 100 years of voting history, from 1913's "Women March for Voting Rights" to 2013's "A Step Backwards," describing the Supreme Court's *Shelby County* Voting Rights Act decision. More than 100 excited and casually dressed teenagers and twentysomethings milled about in a grassy courtyard filled with picnic tables, a beanbag game with an Eighteen × 18 logo, and a small tent to create voting-inspired pins.

Delegates came from all fifty states and were selected because they'd shown motivation and promise as activists already influencing their communities. Among them was a young woman who'd been a page at the Iowa Senate, a Colorado man who, at age twenty-one in 2019, become one of the youngest ever elected to the Denver school board, a female Somali refugee living in Utah, and teenager Auli'i Cravalho, the voice of Disney's *Moana* and a native Hawaiian concerned about climate change.

Delegates laughed as they jumped into an enclosed yellow slide that delivered them from the second floor to an airy lobby. At the bottom were mock voting booths with "ballots" asking which issues the participants were most passionate about and how strongly they identified with a political party (though it didn't ask which one). Hanging throughout the space were signs with voting-related vocabulary, like one for "swing states" above floating crocheted swings.

"When you look at voting as an actual activity, oftentimes it's created as an upper-middle-class hobby. It's [something you do] if you have the time—if you don't have an hourly job, if you have the ability to stand in line and that time works out for you, if your family isn't dependent on your presence somewhere else," Shahidi told the group in her opening address.

She urged her peers to understand that policy debates are discussions about issues directly affecting their lives, and that young people should influence those discussions. "This is a moment to be in conversation with one another so that we can figure out what matters to us," she said. "It should be what matters to every elected official."

She wanted the attendees to take their ideas home and spread the enthusiasm among their peers. After a couple of hours of brainstorming in small groups, each group presented its ideas to snaps of approval from fellow attendees, as the picnic tables were being adorned with flowers and candles for dinner. The groups described how they hoped to grow the youth vote by providing a bus to transport people to polling locations in states where citizens can register on Election Day, by asking YouTube stars to integrate voting explainers into their video content, and by building a website to match people with a "buddy" in their city so each could hold the other accountable for going to the polls.

Meeting potential voters where they are was a big area of concern for the delegates, who feel that the young are often forced to find their own way. They talked about staffing voter registration tables at concerts, and even briefly entertained a pie-in-the-sky idea of an app that summons an authorized person to come record someone's vote. (This is the Postmates generation; they can usually get what they want when they want it. West Virginia in 2018 allowed military personnel overseas to vote via a secure app that confirmed their identity through photo recognition, so maybe it's not so far-fetched.)

One group brought up perhaps the most practical solution of all: create a more informed young electorate by increasing civics education and bringing registration information to eighteen-year-olds and soon-to-be eighteen-year-olds where they're most easily found: high school.

Go Where the Kids Are

While high schools are an obvious place to register young voters, it's far from standard for high school students to have the registration process explained to them, or for them to be taught the mechanics of voting. That's true even in the states that allow sixteen- and seventeen-year-olds to preregister so their registration is automatically active when they turn eighteen. (The District of Columbia does, too.)

As with all things in voting, some states are better than others in encouraging schools to register voters. Various nonpartisan organizations large and small are trying to fill the education gaps and crack the code of registering would-be voters as early as possible. One focus is promoting peer-to-peer registration—training high schoolers to help register their classmates.

Since 2014, Inspire2Vote (previously known as Inspire U.S.) has done that work in states including Arizona, Colorado, Kentucky, and Virginia, with plans to expand as the 2020 election nears. Inspire works with groups that already have contacts in the community to identify students and teachers or administrators who would be receptive to the training. Both Inspire and The Civics Center, another peer-to-peer organization that focuses on preregistration, described similar goals: creating programs at schools that are sustainable and carry on from year to year, with the leadership transferred from student to student, like student council positions.

Andaya Sugayan, who graduated from college in 2017, works in conjunction with Inspire and other civics-related charities to run, as a one-woman band, Inspire PA. She's in her car constantly, driving between high schools in and around Pittsburgh and Philadelphia to register young voters, providing additional voting pledge cards when classes have run out, and answering, via text, questions about primary elections from students and teachers alike. Sugayan, who grew up near Seattle, is high-energy and exudes capability, though her youthful appearance and high-top sneakers make her indistinguishable from the students. She credits her high school government teacher with inspiring her to do this work, recalling him holding mock elections with whatever candidates were on the next ballot, and cajoling surrogates for those campaigns to visit the school and pitch their platforms.

So Sugayan was delighted last fall when a school she visited to help register students also held their student council elections on the same day, and had local election officials bring the city of Philadelphia's new voting machines to use in the election.

Sugayan's style of get-out-the-vote work can be found in the bios of many successful public servants. In her book *Becoming*, Michelle Obama describes how Barack Obama was asked to run the Illinois chapter of a nonpartisan voter registration organization when he was in his early thirties. "To say that Barack threw himself into the job would be an understatement. The goal of Project VOTE! was to sign up new Illinois voters at a staggering pace of ten thousand per week," she wrote, adding that the future president said at the time that the hardest people to reach were—you guessed it—the eighteen- to thirty-year-olds.

Young people often get pegged as apathetic, but those who work with high schoolers find that teenagers simply haven't yet learned to connect voting with its potential effect on their lives. At many Inspire school visits, students ask questions about aspects of registration and the process of casting a ballot that make it "very obvious that we are the very first people talking to the students about the voting process," Inspire's Mixdorf said. They discuss both the mechanics of voting and how politics has a direct impact on students' lives by focusing on relevant issues like how school board decisions affect school facilities and extracurricular activities.

Making registration a team sport of sorts has also proven an effective tool in the success of peer-to-peer registration. At one West Virginia high school, student leaders registered 100 percent of their eligible student body and then helped start a voting program at their rival school. Both schools bused students to the polls for the 2018 primary elections. Also on the bus? A middle-aged teacher who registered with her students and was also voting for the first time. Laura Brill, founder of The Civics Center, told of a Southern California high school that registered or preregistered 630 students. A rival high school said they could do better, and bested them by more than a hundred registrants. Multiple secretary of state offices have gotten in on the competition vibe, giving awards to schools that do a particularly good job registering students.

The Civics Center launched in California in September 2018, and had helped facilitate or provide information for high school voter registration drives in at least twenty-five states a year later. Brill, who said many students and teachers approach her for help after finding The Civics Center on Instagram, provides a step-by-step guide for holding a drive, and suggests reaching out to local organizations, like the League of Women Voters, to help out.

The Civics Center sends teacher contacts a "Democracy in a Box," including stickers and streamers to celebrate the drive, a clipboard with the planning worksheet and state-specific information, including voter ID requirements, plus posters to hang up to publicize the event. The box also includes candy in a "Democracy in a Bag" tote. As is practically required by YouTube-fluent young people, The Civics Center posted an "unboxing" video to show it all off.

Students are encouraged to ask their classmates to list a contact number or email and three friends or family members they promise to get to vote. The cards allow students who are too young to preregister to participate, and give The Civics Center a way to contact them around election days and to remind them to nudge the three people they selected.

Inspire also makes it a priority to follow up with students who've registered or pledged, and text messages are sent in waves as an election nears. Before the registration deadline, they remind students to confirm their registration is current and tell them about early voting. Two weeks before an election they urge voters to research what's on their ballot and provide a link to check their polling location. The final text is on Election Day. In 2018, turnout for the youngest eligible voters, ages eighteen to nineteen, was only 23 percent nationwide; the turnout for the roughly twenty-five thousand eighteen- and nineteen-year-olds who registered or pledged to vote through Inspire programs was 41 percent, the organization's tracking showed.

It's both the indication that peer-to-peer registration works and the huge opportunity for growth that inspires Civics Center founder Brill, an attorney and a former law clerk for Ruth Bader Ginsburg. Brill was surprised to learn, in 2017, that sixteen-year-olds (her daughter was sixteen at the time) could preregister in California. She discovered that hardly anyone she asked knew, either. She wanted to do everything she could to get the word out, and explored how other states were using similar laws. She gathered data and discovered that less than 40 percent of eligible teenagers were preregistered in each of the states that have the preregistration option, and in many of those states the percentage was much lower.

Schools need training and support, Brill said, but the true power comes from the students. Student-driven events allay administrators' concerns about partisanship from outside groups. "When it's the kids wanting what's fundamentally an educational experience about civics and voting, it's very authentic," she said.

TAKE A CHILD TO VOTE

Voting is learned behavior. If a parent or other influential adult discussed voting with you or took you to vote, you're later more likely to be a voter yourself. My mom let me tag along on Election Day. We parked in our town library's circular drive and I'd steel myself for the arctic blast of Texas air-conditioning. Mom would chat with the poll workers, I'd listen closely for any good gossip, and then she'd fill out her ballot. In the humdrum, she taught me to be a voter.

Take your child or favorite niece or teenage brother with you to vote, and explain the mechanics while you're at it.

THE DAY BEFORE

- Tell them they're joining you to vote and, if necessary, what voting is. Build it up as the exciting thing it is!
- Explain that you have to register to vote, and look up when they'll be eligible. Some states allow sixteen- and seventeen-year-olds to preregister and even vote in primaries if they'll be eighteen by Election Day.
- Tell them about the first time you voted, and explain why voting is important to you.
- Show them news articles about the election, and talk about what races are on the ballot.
- Discuss differences between candidates' views in an age-appropriate way.

- Explain which ID, if any, you'll need to bring with you in order to be able to vote.
- Be clear that voting lines are sometimes long. Have them bring a book or activity to help pass the time.
- Check the weather! Will you need an umbrella? Sunglasses? A coat?
- Mention the reward for voting: A sticker! Look up your local "I Voted" sticker.

ON THE WAY TO VOTE

- Explain that almost all adults have the right to vote now, but that it wasn't always the case. Tell them when you would have first been able to vote.
- At check-in, explain what the poll workers are doing—confirming your address and that you're in the right place.
- Have the child say hello to a poll worker and ask a question. It will let the child know that poll workers are there to help.
- As you fill out your ballot, talk about why we get privacy to vote.
- Talk about the importance of reading everything and choosing one box per race. If it's a paper ballot, note that you're filling in the ovals fully and neatly.
- With a poll worker's permission, let the child file the ballot or feed it into the machine.
- Don't forget the sticker!

THAT NIGHT OR THE NEXT DAY

- Check election results and discuss who won and lost, and by how much.

- Some of these tips may need to be modified if you vote by mail, but the idea is the same—include children in the process and discuss the steps in detail.

Reaching the Underserved

Meeting people where they are isn't an idea just for high schoolers. It can also be effective to reach low-income Americans, a group with a history of low turnout. Households earning about $20,000 to $30,000 had turnout rates of about 50 percent in 2016, 30 percent lower than the turnout rate of the wealthiest American households.

Nonprofit VOTE, a Massachusetts-based nonpartisan group, works with nonprofit associations and civic engagement initiatives nationwide to help them register and educate voters who use their services. Participating organizations included community health centers, food pantries, and housing associations.

In 2018, the group registered or received voting pledges from more than 22,500 people at sixty-four nonprofit sites; their success in getting those people to vote indicates the broad potential for the program. The voters engaged by Nonprofit VOTE partners were more than two times more likely to be non-white and under twenty-five, and nearly two times more likely to be in households earning less than $30,000 per year, than were other registered voters in their states.

At a community organization in rural Divide, Colorado, programs include parenting classes, helping families create a budget, and enrichment classes such as those that offer preparation for the General Educational Development test, or GED. (Passing the GED provides the equivalent of a high school diploma.) When circumstances allowed, staff members discussed registration and voting in client interactions. There were home visits to new parents, and cooking classes were offered in conjunction with a food program.

For some clients, it was the first time they'd been presented with the option of registering. "Wow, nobody has ever asked me about that before" was a common reaction from clients, said Kathy Cefus, deputy director of the Community Partnership Family Resource Center. "It was a pleasant, surprise-type reaction." Others expressed concern their vote wouldn't matter, and Cefus found it helpful to mention local ballot initiatives where the margins of victory are small.

Of the people whom Nonprofit VOTE either registered to vote or pledged to vote, turnout was 11 points higher than among comparable voters, with young people aged eighteen to twenty-four voting at a rate 20 percent higher than their counterparts; voters with household income levels of under $30,000 and $30,000–$50,000 both had a turnout rate of 14 percent higher than similarly situated registered voters in those income brackets.

While investment of time is of course necessary, the voter drive programs took little funding and mostly required only staff or volunteers already working with the organization, said Nonprofit VOTE research and field coordinator Caroline Mak. Many groups were able to implement a voting initiative by learning from Nonprofit VOTE's online tool kits.

The nonprofits approached registering and seeking voting pledges from clients in various ways, including asking people as they completed initial intake paperwork if they were registered or would like to be. They set up tables in lobbies and at community events, or visited classes held by the organizations.

Nonprofits are particularly well placed to contact underserved voting communities because they've already built trust, Mak said. They might know to visit a particular transit stop with a lot of foot traffic. They also reach people for whom voting has taken a back seat to a personal crisis, she said. But even in those circumstances, if service providers brought it up, clients were generally receptive.

VOTING TIP: Nonprofit VOTE partners learned to ask the right question: "Are you registered to vote at your current address?" or "Have you moved since you last voted?" was more effective than just "Are you registered to vote?" If you've moved, you need to update your address.

Jerome Sader helped lead voter registration efforts for Housing Action Illinois, an advocacy group that works with member organizations throughout the state serving the homeless and those at risk to be. According to Sader, securing registration and pledges to vote has benefits beyond empowering the individual. "When the people that we serve are voting, it signals to elected officials that people experiencing homelessness are a significant constituency," he said. Like any other group with a low turnout rate, the more underprivileged communities can build turnout, the harder they are for politicians to ignore.

Positive Peer Pressure

Michelle Obama's When We All Vote is trying to get newly eligible voters and historical non-voters to the polls through a mix of education, community outreach, and friendly peer pressure. Obama is able to harness major star power—the group's debut announcement included Lin-Manuel Miranda, Chris Paul, Janelle Monáe, Tom Hanks, Faith Hill, and Tim McGraw. But among the most important things they did, according to Kyle Lierman, the CEO of When We All Vote, was establish a hashtag that would encourage everyone—from Cardi B to Iowa teenagers—to vote and to bring their friends along.

Obama herself kicked it off, asking people to choose five friends or family members to be in their "#VotingSquad," and then to make sure each voted. "You are the best messengers to get out the vote. You know who's too busy or too forgetful or who might flake out on election day . . . Tag each other on Instagram and Facebook . . . put everyone on a text chain, and then get to the polls," she said. It resulted in a seemingly endless scroll on social media of #VotingSquad selfies, including parents escorting their twin daughters to cast their first votes and women contorted into yoga positions, their bodies spelling "Vote."

Obama's organization is also nonpartisan, and one of their first priorities was to go where the voters aren't. They focused digital ads and social media pushes in zip codes with low registration rates and recruited volunteers in those communities. In 2018, they organized 2,500 local voter registration events. Going into 2020, they've asked volunteers to be "squad captains" who receive training on recruiting voters. They are given suggestions for approaching different targets each month, such as faith groups, to register new voters. When primary elections were postponed or altered due to the coronavirus pandemic, the group kept people up to date on social media on the status of the elections and options to vote by mail.

Stumping for Their Team Only

Other Barack Obama administrative alumni are in the get-out-the-vote game in a way that's unabashedly for their own team. Former speechwriters Jon Favreau and Jon Lovett and former National Security Council spokesman Tommy Vietor founded Crooked Media in 2017. They have a variety of podcasts and a news site but are best known for *Pod Save America*, which reached its first million-listener episode shortly after its launch.

It covers the news from a left-leaning perspective and frequently welcomes Democratic candidates and commentators. The founders' goal was to avoid jargon and instead to talk about politics like real people do. When I first interviewed Favreau, the podcast business was so new he was personally handling all media requests and scheduling.

Fast-forward two years to a late winter visit to their West Hollywood offices, a loft-like floor in a commercial office building where visitors are greeted by a huge neon outline of George Washington's head, white headphones in his ears, his image backgrounded by wallpaper composed of newspaper pages, a few from the Watergate era. Favreau's caramel-colored dog played with Vietor's chocolate-colored one in the corner office.

Shaniqua McClendon, Crooked Media's political director, sat nearby. She joined the company after graduating from Harvard's Kennedy School of Government to help lead its "Vote Save America" effort. The company created the position to help ensure that their audience of engaged listeners were also engaged voters. Going into 2020, Vote Save America took up fighting gerrymandering and partnered with Stacey Abrams's Fair Fight 2020 to raise money for voter education and to fight voter suppression and educate voters. They raised more than $1 million from listeners in just over two weeks.

Voting mechanics and voter suppression have been top of mind since the 2018 midterms. During the podcast, listeners are regularly encouraged to visit the Vote Save America site to register or confirm their registration. ("Sometimes names get removed by mistake or by people who hate, what's the word, democracy," the Vote Save America registration page said.) The company implored listeners to volunteer and connected them to campaigns in need. They signed people up for 23,000 volunteer shifts for Democratic candidates across the country, a number that felt low to McClendon until she remembered they'd had only about two months to do it. As with When We All Vote's #VotingSquad, voters and volunteers used a hashtag, #VoteSaveAmerica, to share their experiences. Before the November 2019 elections, Vote Save America reminded followers on Instagram to post their volunteer efforts, saying, "Peer pressure! Sometimes it's good."

Pod Save America had good reason to focus on canvassing. A taped segment that ran during their HBO show featured Vietor, Lovett, and comedian Akilah Hughes knocking on doors in a California neighborhood. The tone was a mix of sarcasm, silly humor, and earnestness. "Canvassing doesn't just turn out the people you talk to. The other people in that home are also sixty percent more likely to vote," Lovett said at the end of the video. "So maybe put down your phone, stop tweeting for two {expletive} seconds, and actually hit the pavement. Because it's the single best thing you can do."

Volunteerism was the touchstone for Swing Left, another (just like it sounds!) left-leaning organization that threw its energies into harnessing volunteers to work for races where conservatives were vulnerable. They wanted to switch certain House of Representatives seats held by Republicans to being Democratic ones. Swing Left's founders came from outside politics, and their idea was a unique one—raise money in specific districts to be used for whichever candidate won the primary, and help people find their closest volunteer opportunity.

On a Thursday night in Manhattan, the last before the 2018 midterm elections, a line snaked around Cooper Union's Great Hall as Swing Left volunteers waited to get into the building where Abraham Lincoln and Elizabeth Cady Stanton both once spoke (separately, of course).

Backstage was Zoe Petrak, a Fordham University senior and Swing Left college fellow. She was camera-ready in a white V-neck t-shirt and black leather jacket, ready to take the stage with Ruffalo, the *Avengers* actor, to explain why she'd volunteered. "The 2016 election was the first election I was eligible to vote in," she told the crowd, but afterward she'd regretted not finding additional ways to participate. "I didn't do enough! I just voted."

When she faced indifference from her peers, she used volunteerism as a way to extend the power of their single vote. "If you go out and talk to people . . . that's how you make your vote matter more," she said.

Fieldwork can't flip seats that are polling 60 percent for one candidate and only 40 percent for the other, said Adrienne Lever, who served as Swing Left's political director. "But it can do it when it's a lot tighter." Lever considered about thirty-five House races to be toss-ups going into the midterms, and said that the "national political dynamic" was causing certain races to be closer than expected. "But they're going to get across the finish line because of the work of people like Zoe."

Canvassing certainly isn't an activity limited to those on the left. In 2018, conservative Ralph Reed's Faith and Freedom Coalition reported using information mapping software and "a smartphone app that locates faith-based voters" to knock on doors in twenty-one states; they targeted people with historically high voter turnout rates, or those who self-identified as evangelicals or Roman Catholics.

While those reached by Reed's group were likely pleased the Republicans maintained their majority in the Senate, the Democrats surpassed the twenty-three seats they needed to become the majority party in the House. Overall, Democrats gained forty-one seats.

The point of highlighting the work of organizations like these is not to tell you how to vote but to show how some of the large, coordinated efforts are working to get out the vote, educate others, and spread the message. The most successful drives involve making the message personal, whether it's knocking on doors or recruiting your friends.

Make It Personal

Perhaps no get-out-the-vote movement has felt more personal than that of the students and alumni of Stoneman Douglas High School who founded March for Our Lives and oversaw its voter recruitment efforts.

Several March for Our Lives members traveled to Nashville to emcee the event where Sheryl Crow was performing—a pre-midterm morning concert and voting party in Nashville organized by a fellow alumnus. Ramon Contreras, who lost a friend to gun violence in Harlem, also represented March for Our Lives that day. Sitting around a table backstage, participants described how many of their conversations with fellow citizens were about finding common ground, and how they told people that the best way to make their feelings known to pubic officials is to vote.

Apathy, they've found, can be combated with simple logistics. "Sometimes it's just as easy as making a plan with someone. They don't know where their polling place is, they don't know [poll] hours . . . But if you go through it with them and show them the resources and walk them through a solid, tangible process, they're much more likely to actually vote," said Brendan Duff, a 2016 graduate of Stoneman Douglas. His brother Daniel, also at the table that morning, was a freshman at the time of the shooting.

Several of the Parkland kids wore white sweatshirts with an American flag, but in place of the rows of stars was a QR code that can be scanned with a smartphone to lead directly to voter registration sites. They'd done an organized "shirt drop" at over a thousand schools and colleges, said Matt Deitsch, the organization's chief strategist and another alumnus.

The group displayed a mix of awe that they were on such a whirlwind ride and raw optimism that what they were doing might eventually work. Their high school's namesake would likely have been impressed with their efforts. Marjory Stoneman Douglas died in 1998 at age 108 and was awarded the Presidential Medal of Freedom in 1993 for her environmental activism. But in her twenties, she had another cause—fighting for the women's suffrage amendment.

The youngest people at the table that day in Nashville, Kirsten McConnell, a Parkland senior, and Daniel Duff, were quiet for most of the discussion. But when asked if they had anything to add, Daniel, several years shy of voting age, laughed a little, and joined the chorus of others sharing a single message: Vote!

CHAPTER FIVE

MAKING VOTING THEIR BUSINESS

Patagonia offices in Ventura, California, November 2018 (Tim Davis/Courtesy of Patagonia)

The Party at the Polls that brought the March for Our Lives members to the Nashville amphitheater was a morning combination of concert and civic duty. With the city skyline as a backdrop, Sheryl Crow surprised the crowd with a duet of "Every Day Is a Winding Road" with country "it girl" Maren Morris and encouraged everyone to bring a friend to the polls. Billy Ray Cyrus strummed his guitar in a darkened dressing room before he went on, speaking with genuine concern about the North Dakota ID law causing problems for Native Americans preparing to vote. Grammy Award winner Jason Isbell performed a song about Americans' shared burdens and fate with his wife, The Highwomen singer and violinist Amanda Shires. Shortly after, the two led the crowd of about 1,000 walking to early vote.

Most of the performing artists were represented by WME, a subsidiary of the talent, entertainment, and media company Endeavor. Additional clients include Rihanna, Emma Stone, and Dwayne Johnson. Marissa Smith, then an agent-in-training in WME's Nashville office and an alumna of Stoneman Douglas, was the strategist of the event. She wanted to use her employer's star power to promote voting, and her idea was a straightforward one—gather A-list musicians to perform and then immediately escort the audience to the polls.

Endeavor is one of hundreds of companies involved in organized efforts to promote voting, including encouraging their own employees to vote and holding registration drives for their fans and customers. Businesses have multiple reasons to encourage voting. Being good corporate citizens and promoting democratic ideals is certainly part of it, but corporations also see business benefits. A 2019 Harvard Kennedy School case study on the power of companies to increase voter turnout found that the business positives included escalated employee satisfaction; pleasing customer bases, whose constituents increasingly expect corporations to be civically responsible; and heightened brand awareness.

Supporting democracy is also important for the very existence of capitalism, said Eric Orts, a professor of legal studies and business ethics at Wharton at the University of Pennsylvania. "The whole basis of a free enterprise system assumes a democracy," he said, and failing to protect that democracy can only spell bad news for the corporate world. The most straightforward way for businesses to do their part is fairly simple. They should make clear they're not taking sides but that they support democracy and an active electorate.

"We need a culture shift around voting, not because of politics, but because it is this fundamental American right that we need to take seriously," said Corley Kenna, head of communications for outdoor clothing company Patagonia. Patagonia led a coalition of businesses that pledged to give employees time off to vote in the November 2018 election. Company leaders, especially if their numbers reach critical mass, are in a great position to drive home the idea that voting should be an automatic action for responsible citizens, she said.

Endeavor's Voting Adventures

At the Endeavor Party at the Polls, the blaring brass instruments of a second line band provided a festival feel as the crowd took the ten-minute walk on a closed downtown street from the amphitheater to the polling place. The marchers carried serious signs ("Engage Connect Empower") and funny ones ("Taylor Swift, Look What You Made Me Do"), the latter referring to the pop star's having implored her fellow Tennesseans to vote. Some parents guided their young children along for the civic outing, while fellow marcher Barry Towles joined his beaming eighteen-year-old daughter, Aerial, as she walked toward her very first vote. "I'm extremely excited," Aerial, who is African American, said. "I'm really inspired by all the sacrifices that have been made for me to be able to vote." The stories she heard growing up about her own grandfather's suffrage struggles made the moment particularly special.

Marissa Smith, the event organizer, was overjoyed her employer had supported her idea. "This is my way of healing, my way of taking action," she said. It's nearly impossible to pull off an event like that without the backing and resources of a large organization like Endeavor.

The Party at the Polls was only one way in which Endeavor experimented with promoting voting. The company believes that businesses can no longer sit on the sidelines and fail to actively promote civic participation, said Amos Buhai, head of Endeavor's government relations group. Endeavor also had a little healthy competition from their fellow big shot talent agencies. The *Hollywood Reporter* gave the rundown: United Talent Agency was hosting phone banks and voting parties in their office, and letting employees take corporate ride-shares to vote; CAA provided ballot-specific information to employees and played a large role in supporting and promoting a voting initiative called "I am a voter"; ICM Partners hosted a live-stream event with celebrities including Julia Louis-Dreyfus and Chelsea Handler to promote turnout, and prepared a voting guide for employees.

For companies expanding their voting initiatives, the divisive and often unpredictable political climate gave them incentive to take a little corporate control of the situation. Endeavor's corporate officers, for instance, started a PAC to promote progressive causes, and their charitable component took on a gun-control organization as a client after the October 2017 mass shooting in Las Vegas, when one of their artists was onstage and employees were on-site. But whether companies' reputations are left- or right-leaning, those interviewed were adamant about keeping their get-out-the-vote efforts nonpartisan. They ask only that people vote; they don't suggest whom they should vote for. Companies, especially large, national ones, have employees, customers, and partners from both parties and don't want to alienate any of them.

A few months after the Party at the Polls, at Endeavor's headquarters overlooking the luxury stores and palm trees of Beverly

Hills, Buhai talked about the company's other voting-promotion efforts. They partnered with Lyft and Twitter to support the push for automatic voter registration in Maryland, New Jersey, and Nevada. (AVR was approved by popular vote or gained legislative approval in all three states.) Their goal went beyond just getting AVR passed; they also wanted to publicly demonstrate their belief that "participating in the electoral process and participating in civic engagement is ultimately good for business," Buhai said.

Internally, the company—which is also the parent company to Frieze art events and festivals and Ultimate Fighting Championship, among others—set a lofty goal of 100 percent employee voter registration, and mined voting and secretary of state websites to inform their employees across the country of registration dates and early voting opportunities.

Going into 2020, Endeavor is improving on that model by working with employee volunteers in their different locations and subsidiaries to educate employees on local registration periods and early voting opportunities. The volunteers, Buhai said, come from various departments and levels, and include a UFC employee, a coordinator of TV packaging, and an executive at Endeavor Content.

Endeavor wants to make it as convenient as possible for employees to vote. They brought attention not only to the 2018 midterms but also to primaries. Employees nationwide were given paid time off to participate in their primary or for early voting, and were reminded when the primaries were. It was made clear that time off to vote truly meant time off. "In this environment there are a lot of younger people who are assistants, people who work crazy hours," said Marie Sheehy, a communications executive at Endeavor. "Managers were told, 'You gotta let them go.'"

Time to Vote × 400

In June 2018, Patagonia CEO Rose Marcario wrote a public blog post saying that the company would close all of its offices and stores nationwide for the midterm elections, as it had for the presidential election in 2016. Citizenship, including corporate citizenship, requires supporting democracy, she wrote.

Companies immediately reached out to Patagonia, asking for advice on doing something similar. Corley Kenna, the communications head, saw an opportunity to create a coalition and make the enterprise more visible. She had worked on Capitol Hill and for Hillary Clinton's State Department, as well as in corporate public relations for Ralph Lauren, so this was right up her alley.

Many states have laws requiring employers to give employees time off to vote, but it's usually a limited amount that may not work if you face long lines or a polling place far from work. And if employers specifically tell employees they have the time, it helps remind them to take it.

If Endeavor's LA space is urban modern, Patagonia's just-off-the-beach offices are homespun ski lodge, a small collection of earth-tone two- and three-story buildings, with the tin shed where the company was founded still standing nearby. The chairs in the cafeteria are covered in delightfully messy murals painted by employees' children; their playground is next to the pergola under which adults eat lunch. A basket of there-for-the-taking oranges sat near a front entrance.

Over lunch in not-playground-adjacent downtown Ventura, Kenna discussed how the different companies handled the "Time to Vote" logistics and how Patagonia hopes to expand the effort and impact. "We didn't have a formal outreach strategy, nor did we have a budget," Kenna said. "Over the course of three or four months, we got 400-plus companies." Patagonia's only expenditure was a one-page ad in the *New York Times* on National Voter Registration Day, asking other companies to join the effort.

VOTING TIP: The Harvard civic responsibility study found that companies can do a lot to encourage voting without spending much money, and that includes companies with robust, decades-old get-out-the-vote initiatives, like Blue Cross Blue Shield. Companies said worthwhile expenditures were "swag activations" such as magnets featuring the Election Day date or a photo booth for employees to pose with their "I Voted" stickers.

Patagonia is known nearly as much for its environmental activism as it is for its fleeces, and even endorsed two pro-environment candidates in 2018. In part because of that reputation, Kenna and Patagonia thought it important that the initiative include companies with diverse customer bases and geographic locations. They wanted Time to Vote to focus purely on civic action, and for people to believe the companies involved meant it. Kenna was pleased to have participation from corporations like Arkansas-based Walmart and Tyson Foods, owner of household brands including Sara Lee and Jimmy Dean, because they reach a politically diverse client base throughout the nation.

Those businesses and others, including Nordstrom and PayPal, also checked another desirable box. "People identify with those companies," Kenna said. It helps drive home the importance of doing your civic duty when brands that citizens respect—and even consider part of their lives—dedicate resources to promoting voting. Smaller companies signed on as well. Massachusetts-based

Prosperity Candle sent fellow participants inspiring notes about why they felt the project was integral to democracy.

Moving into 2020, the initiative has grown in both size (their goal is 1,000 companies participating) and purpose, keeping the "time off" portion but also encouraging participating businesses to actively discuss the importance of voting as a civic responsibility. Some in the first group of four hundred businesses had already moved beyond just time off, including Walmart, which has a dedicated website that helps employees register to vote and find their polling place. Levi Strauss & Co.'s many voting-related activities included a commercial of a diverse group of people casting their ballots around the world, to a soundtrack of Aretha Franklin's "Think." "Levi's is all about authentic self-expression," chief marketing officer Jen Sey told *AdAge*. "And there is no purer form of self-expression than voting."

HOW EMPLOYERS CAN ENCOURAGE VOTING

- Set the tone from the top: Send an email or post video of executives discussing their plans to vote.
- Appoint a point person (or point people, depending on company size) to lead the efforts, and check local voting laws and relevant dates.
- Set a goal of 100 percent voter registration by eligible employees. Notify employees of voter registration deadlines, and explain how they can register. Ask employees already registered to sign a pledge to vote.
- Talk to a local voting organization about hosting an on-site registration drive.

- Consider friendly competitions between departments (or even between competing companies) over who can gather the greatest number of registrations and pledges.
- Inform employees in a timely manner about state and local voting options—early voting, absentee, vote by mail—and what ID they'll need, if any, to vote.
- Suggest that employees sign up for a voting service, like TurboVote, to help them easily find their polling place and be reminded of upcoming elections.
- Offer employees time off to vote, and mean it!
- Make Election Day meeting-free, or at least limit meetings to truly time-sensitive ones. It allows more people to work from home and closer to their polling place.
- Provide information about what's on your local ballot, and bring in experts to discuss, in a nonpartisan way, issues that might be confusing to voters or relevant to your business.
- Invite candidates whose decisions would affect your business to speak to employees.
- Create company-specific voting swag—stickers or magnets. "(Your Company Name Here) Votes!"
- Brag! Show off your voting efforts on social media, both for registration and on Election Day.

What Employers Can Do On-Site

As some workers left their shifts and others arrived for new ones at the gold mines in Teller County, Colorado, in late September, they were given the opportunity to register or pledge to vote. Their employer, Newmont Goldcorp, had invited the Community Partnership Family Resource Center to visit the job site as part of National Voter Registration Day after executives learned about the voting work the center had done with its clients. It's an example of a simple, one-day event companies can do to increase registration and turnout among employees.

An organization that spreads that message more broadly is the Iowa Association of Business and Industry, which serves more than 1,500 member companies from the state's 99 counties. Part of the service that's provided is encouraging those companies' 330,000 employees to vote. The organization's Iowan roots date to 1903; it is the state's answer to a Chamber of Commerce. They help facilitate and provide information for voter registration drives, send posters with voting information to hang in break rooms, and even provide paper reminders with paychecks for those who haven't joined the world of direct deposit. On Election Day, the Iowa Association sends links to help members' employees find their polling place, know what to do if they want to register same-day (an option in Iowa), and what ID they'll need once they get there.

A lot of the Iowa Association's information on how to engage employees comes from the Business-Industry Political Action Committee (BIPAC), said Nicole Crain, the Iowa Association's executive vice president. BIPAC's stated mission is educating, motivating, and activating employees to participate in the political process.

In a study BIPAC commissioned surveying public sector employees after the 2018 election, 61 percent of people who received voting information from their employer said it made them more likely to vote. Employees "want to hear from their employer on elections and politics; and overwhelmingly find the information provided to them to be helpful when it's shared in a non-partisan and objective way," BIPAC CEO Jim Gerlach, a former Republican U.S. congressman for Pennsylvania, said in the report.

Still, some companies are squeamish about dipping a toe into the election waters, fearing controversy and even legal trouble. The association tries to calm concerns by explaining it's completely okay to provide voting information and to encourage civic participation. Just steer clear of telling people how to vote, Crain said. Nerves are also eased by encouraging safety in numbers. At an annual meeting of member companies, the association gives a legislative update on new laws that affect businesses and talks about their get-out-the-vote resources, encouraging hesitant businesses to speak to companies in attendance that have active programs.

Personal Voting Awakenings: Pass Them On

Clique Brands has learned that its diverse client base has mixed feelings about them stepping into anything political. The media and fashion company reaches an audience of more than twenty million a month through its website, Who What Wear, and its newsletter and social platforms. The company also has a popular women's clothing line at Target and a widely available skincare line, Versed.

After Who What Wear posted about the 2017 Women's March on social media, co-founder Hillary Kerr said, "We had incredible engagement, but it was all over the place." Many appreciated that they were covering it, but others were angry—the company heard a lot of "Stay in your lane." As a result, they had to decide when and how to wade into politics. Encouraging voting hit the sweet spot of nonpartisan civic engagement.

It felt both of paramount importance and doable from a business perspective. Customers and readers were receptive to the idea that voting is a responsibility and gives them some say over how the country is run. "It dovetails nicely with this idea of being CEO of your own life," Kerr said, an idea Clique uses to encourage young women to feel autonomous and to remind them that their voice matters.

For the 2018 midterms, Clique worked with the group "I am a voter" to institute an hour-long "digital shutdown" across their sites on National Voter Registration Day. Companies participating in a digital shutdown take a break from posting new content, instead encouraging readers visiting their sites or social media to take the time to register to vote; they provided a link to a voter registration site. (The day before the midterms, @WhoWhatWear posted for their more than 2 million Instagram followers a photo of actress Tracee Ellis Ross wearing a red ball gown skirt and an "I am a voter" t-shirt, captioned "Heading into tomorrow like . . . #VOTE.")

Like Endeavor and Patagonia, the company's approach to voting involves both internal and public-facing action. Kerr is a somewhat recent convert to voting evangelism—she readily discloses that she was not very politically active until a few years ago, and said learning to educate herself about voting and politics at times felt overwhelming. In hopes of encouraging younger employees to avoid that same hesitation, she tells them that her door is open to answering questions, and that she wants to help voters think through issues and find information. That is in addition, she said, to her friendly but incessant voting reminders. "I don't care how you vote. I don't care when you vote . . . But it's really important."

Kerr was refreshingly straightforward about the logistics of providing flexibility to facilitate taking time to vote. "There was great collaboration and teamwork on making sure that all pressing business needs be met. But there's no emergency in this office that is so great that it cannot withstand a few hours off to go vote," Kerr said.

Employer-Led and Running With It

In 2016, wide receiver Kenny Stills, then with the Miami Dolphins, got interested in politics and social action, and decided his National Football League teammates should register to vote, too. He stuck voter registration forms in their lockers and heard . . . very little. Undeterred, he took his idea up the chain of command, and eventually Dolphins owner Stephen Ross got his organization, called Ross Initiative in Sports for Equality (RISE), involved.

RISE helped with the logistics of registering, and by the end of one practice session in June 2017, every Dolphin had registered to vote. (In an example that two people can support voter registration and strenuously disagree over whom to vote for, Stills publicly criticized Ross for hosting a fundraiser for President Donald Trump in the summer of 2019. Later that summer, Stills was traded to the Houston Texans.)

Following its work with the Dolphins, RISE formed RISE to Vote, and by that autumn had held formal registration sessions with the NFL's Atlanta Falcons and the National Basketball Association's Atlanta Hawks, the Brooklyn Nets, and the San Antonio Spurs. As the organization grew, they worked with Major League Baseball teams as well, and held education and registration events with more than twenty college athletic programs. (If a player is already registered, he or she can pledge to vote.)

RISE also held major activations at the Super Bowl and the NBA All-Star Game, and at the NCAA Final Four, where basketball fans could walk through a timeline of sports and social activism, "interview" players in a mock dressing room, and then register to vote.

When RISE meets with teams, they focus first on the educational component. Someone from a local voter registration organization explains the nuts and bolts of registering and voting absentee—a big issue for traveling athletes.

Younger players are often ambivalent about voting, or admit they don't see its purpose. "A lot of the work we're doing is talking about why is this important, what can be impacted," said Diahann Billings-Burford, RISE's CEO. To that end, RISE invites a current or former athlete to share their own story. Former NFL wide receiver Anquan Boldin talked to student-athletes at his alma mater, Florida State University, about how district attorney elections can affect a community.

While management sometimes made the initial invites, Billings-Burford said it's the motivated players who can make the difference. In 2018, the National Basketball Players Association asked RISE to Vote to have a table at their rookie transition program, an event that occurs annually to help prepare young players for life in the NBA. The table was being ignored until one rookie ran up, excited to register to vote, Billings-Burford said. "He walks away from the registration table, goes into the room, and tells everybody, 'You guys need to get registered! They're right outside!'" The table then hosted a crowd.

CHAPTER SIX

THANK YOU FOR VOTING

Anderson Traugott pointing to an "I am a voter" sign in Santa Monica, California, October 20, 2018 (Courtesy of "I am a voter.")

The declaration #ididitforthesticker is popular on social media during election season, with people displaying their "I Voted" stickers in a variety of creative ways—on their noses, on their adorable pointy-eared dogs, on their adorable if less pointy-eared babies, and as an opportunity to highlight an ample chest. Plenty of people also take to social media to complain that they voted but somehow didn't get a sticker.

It makes sense. The stickers are pretty great. New York City's "I Voted" sticker for several years showed colorful intersecting lines reminiscent of the subway map, but was switched in 2019 to the skyline to better match the official election website. Las Vegas's also features the Statue of Liberty and the Empire State Building, along with the Eiffel Tower and the Great Pyramid, all recognizable casino landmarks on that city's horizon. "I Voted" appears on a California sticker in thirteen languages, including Thai and Tagalog. Tennessee voters get one in the shape of their state, while the sticker of its southern neighbor is dominated by a pretty Georgia peach.

Alaska received attention for a recent sticker revamp, trading its usual blue-with-yellow-stars constellation—modeled after the state flag—for a Juneau artist's renderings of native animals doing Alaska-ish things, including a walrus on a snowmobile and a moose in a lumberjack-inspired red-checked shirt. But there was a catch. To encourage early voting, Alaska offered the animal stickers to those who voted before Election Day. Those who waited would have only the vintage blue-and-yellow sticker to show for it. According to the *Anchorage Daily News*, early voters in 2018 outpaced the total number of early voters in the last three comparative Alaskan elections. Could it be that voters braved the long early voting lines for the sticker? Or was it just part of the record-breaking early voting turnout nationwide? Or were Alaska's voters motivated by the excitement of this particular governor's race? (The incumbent had dropped out just weeks before Election Day.)

Gronke, the Reed professor, pointed out the difficulty of ever knowing with certainty, outside of individual anecdotes, how a particular factor affects an election. As he explained, you can't change one element in the political system and force everything else to remain static. When a new voting law is put in place, for example, voting advocates mobilize to inform groups that might be affected by it. Turnout might go up even if the law adds a barrier to voting. If early voting is enacted, campaigns will focus on getting supporters to vote early, so knowing how early voting improved overall voter turnout—or didn't—is difficult.

Before all the turnout experts throw my own book at me, please note that this isn't a knock on them but rather recognizing the difficulty of the work they do. To my mind, anything large or small that encourages people to vote is worth a try.

Voter Encouragement Tactics

Columbia University professor Donald Green and Alan S. Gerber wrote *Get Out the Vote: How to Increase Voter Turnout*, the go-to book on the subject of encouraging voter turnout. They describe experimental research on voter encouragement tactics and their effectiveness in increasing turnout, including robocalls (basically no effect), whether canvassing works (yes, it does, but beware the drunken or excessively ideological volunteer), and if including a voting story line in a Spanish-language soap opera might up voter registration among Latinos (modest increase, at best).

Green's office is situated within Columbia's highly regarded School of International and Public Affairs. Green wore unassuming glasses and a neat blazer and spoke in the patient, considered way professors do when they're trying to condense twenty semesters' worth of knowledge into one class-length conversation.

Get Out the Vote describes social-pressure communications as playing on "a basic human drive to win praise and avoid chastisement." In the case of voting, the praise comes from casting a ballot and the scorn from failing to meet one's responsibility to do so. The research on the existence and variety of social-pressure mailers is fascinating.

The idea for the first social-pressure mailer came from a campaign consultant intrigued by the large number of people who claimed, in post-election surveys, that they'd voted—when public records showed that they hadn't. People thought they could get away with it, the consultant assumed, because they believed voting to be a private act and didn't realize that it's possible to verify if someone voted or not. The consultant ran an experiment on his own and found that sharing people's own personal voting history with them was effective in getting them to the polls. He eventually reached out to Green and Gerber to see if they could team up on experiments to confirm if that type of messaging really works.

Green and his fellow researchers sent out four kinds of mailers leading up to an election. One "scolded recipients to do their civic duty." One told people to do their civic duty and also that their participation was being monitored. The third, called the "Self" mailer, noted whether each voter in the household had voted in recent elections and promised to send a future mailing reporting if the household voted in the upcoming election. Last, a "Neighbors" mailer revealed if household members had voted as well as if people on their block had. A control group received no mailers.

The maximum social pressure—talk of your family PLUS comparing you to your neighbors—increased the likelihood of voting by 8 percentage points, a huge jump in the land of turnout tactics and a statistic superior even to results achieved through face-to-face canvassing.

But consider how creepy it is to get a mailer telling you that someone took the time to look up your voting record, compare it to your neighbors, and then hassle you about it. Like many things that come from social pressure, it may make people act, but they may not act in the way you wanted. People could rush to the polls to vote for someone other than the candidate who filled their mailbox with snooping notices. A Wisconsin organization that used the "Neighbors" version received hundreds of complaints and unwelcome inquiries from journalists.

Campaigns are therefore hesitant to use the high-risk method, though efforts to effectively tweak the language continue. Interestingly, we're apparently more immune to being outed as non-voters if campaigns contact us by phone—studies found that using the social-pressure-type messaging via text was less effective, and sometimes not effective at all.

There is, however, a better option, one that leaves behind the risks of social pressure but still garners benefits—Green described it as a "preemptive thank-you note." These flatter the recipient about their past behavior in hopes that words of praise will encourage them to vote in an upcoming election. The thank-you notes are the kinder, gentler way of saying "Yep, we're watching." They suggest to the prospective voter "that what they do is public record, but you're not going to shame them for it, you're going to make them feel good," Green said. They also might get people to the polls without causing "backlash and consternation in the same way that scolding people does."

Even just thanking people for voting without referencing their past history can increase turnout, studies showed. Like nearly everything else in get-out-the-vote land, there is no certainty it will work, but it's an option with little to no downside for campaigns or get-out-the-vote groups. Most people appreciate being appreciated for doing their civic duty.

"I am a voter."

Some get-out-the-vote language methods move beyond encouraging you to pat yourself on the back. In 2011, research conducted by Christopher Bryan and his colleagues showed that tweaking language to ask people shortly before an election if they'll "be a voter" rather than if they'll "vote" can encourage election participation. In other words, voters who are nudged to adopt voting as part of their identities may be more inclined to go to the polls. A voter becomes something that you are, rather than voting being only something you do. People want to see themselves as competent and deserving of social approval. "Being the kind of person who votes may be seen as a way to build and maintain a positive image of the self—to claim a desired and socially valued identity," the researchers state.

While in graduate school more than a decade ago, Bryan, now an assistant professor of behavioral science at the University of Chicago, was fascinated by experiments that showed that people associated traits as being more part of a person's makeup if a noun is used to describe them ("she is a carrot-eater") versus a verb ("she eats as many carrots as she can"). That work inspired him to test if using similar language cues when talking to voters could frame future behavior.

He and his colleagues conducted three separate experiments, one on unregistered voters who still had time to register, and two on registered voters just before an election who hadn't yet voted but still had time to. They used "verb" language and "noun" language when asking participants if they would be voting. Bryan and his team checked the voter rolls to see if participants had registered and voted, and found that those in the "be a voter" category were more likely to have actually voted. The simple upshot is that he would advise campaigns to use this "be a voter" language in the direct lead-up to an election to encourage action.

The study has prominent naysayers, including Green's *Get Out the Vote* co-author Gerber (dean of social science at Yale), who concluded, based on larger-scale follow-up studies, that noun/verb language is unlikely to make a positive difference in voting behavior. (In a follow-up paper, Bryan said that some of Gerber's methods of replicating the research weren't applicable, and that he believes that some of Gerber's results actually support Bryan's conclusions.) Either way, I like the concept of thinking of "being a voter" as part of one's identity. It suggests an important, habitual part of your life is taking action to support the values you would like your community and country to reflect. That idea complements what voting advocates are trying to instill in new voters.

Many voting organizations, including When We All Vote, Inspire, and Eighteen × 18, keep their discussions of voting, voting education, and registration constant, even in 2019, a so-called off election year. The organizations' goals are to demonstrate that there are no off years, and that voting is something Americans should think about as an ongoing part of our lives, not something to consider every four years.

That idea and Bryan's research both came to mind when I saw pictures from a New York fashion show by designer Prabal Gurung. The seat assignments were printed on a white card with a simple phrase in black sans serif letters: "I am a voter." Within weeks, basketball superstar Steph Curry was holding a small black pin with the same phrase up to the camera.

Using language to encourage considering voting as part of personal identity—similar to someone labeling herself an entrepreneur, film buff, or foodie—inspired the "I am a voter" campaign, said Mandana Dayani, its creator and co-founder. Dayani was born in Iran, and a refugee organization helped resettle her family in America when she was a child. She became a corporate lawyer and later served as vice president of Rachel Zoe, Inc., a fashion company.

After the 2016 election and its constant debate over immigration, the importance of encouraging citizens of all experiences and backgrounds to vote took on particular resonance for Dayani. But it took her some time to find where she fit in the get-out-the-vote world.

Conversations with campaign workers and other political operatives led her to a realization. "Politicians were speaking to voters in political terms about politics in political forums," she said. "In my community of entrepreneurs, we speak to people like they're consumers. We try to take anything that we're trying to teach them or sell them or provide them and figure out how we can make it approachable and exciting and like the thing that cannot ever be missed."

What was missing was a voting brand, she said, that appealed to younger adults. One that wasn't preachy but was inclusive and cool. In spring 2018, she asked her friend Tiffany Bensley, who helped shape marketing for luxury brands like The Row, and about a dozen other fashion and creative types to join her in an initiative to focus people on getting themselves to the polls every time there's a voting opportunity.

A sought-after creative director helped Dayani select Neue Helvetica, a simple and recognizable font, and suggested they spread their message with something clear and forceful and with a period at the end. "That's so much more powerful because you're now declaring a statement about yourself," Dayani said. They stayed away from the stereotypical reds, blues, and waving flags of most political messaging in favor of a chic black-and-white look. Included in the @iamavoter Instagram stories were selfies of people wearing one of their t-shirts, hoodies, or pins, as long as the images weren't divisive.

Both the uncomplicated message and the creators' commitment

to being nonpartisan in the campaign encouraged companies to help them spread the word. Entertainment behemoth CAA became a founding partner in monetary and promotional support. (CAA Foundation executive director Natalie Tran is also a cofounder.) *Star Wars* director/producer J. J. Abrams's production company was active as well. For a month, AMC Theatres ran a public service announcement before movies that ended with the "I am a voter" logo and a number to text for registration and voting information. The digital organization also did something analog: the Ad Council donated billboards and radio ads.

The campaign focused on positive and powerful language and facts about voting rather than using shaming language about low voter turnout rates. They pushed this optimistic spin via social media, asking influencers to share the message and offering downloadable versions anyone could post. Options included "This November, millennials will make up the largest voting bloc for the first time."

In 2019, they included in their social media messages information on states with an active election year, like Virginia, and registered voters at gatherings as diverse as the extravaganza that is the Beautycon cosmetics convention (famously described as "Sephora meets Coachella") and at WE Day UN, a youth empowerment day held during the meeting of the United Nations General Assembly. Shortly after 2019 National Voter Registration Day, Steph Curry invited the "I am a voter" team to help register his Golden State Warrior teammates.

The "I am a voter" founders want to make the initiative a lifelong mission, Dayani said. The founders hope they'll continue to run the campaign, even until their preschoolers and grade schoolers—currently running around in "I am a (future) voter" t-shirts—are grown and working alongside them.

Make It a Party

Success in driving turnout on Election Day can be enhanced by combining voting with an event people want to attend. "I am a voter" had tool kits with resources corporations could use to boost employee participation, including telling them about "#VoteTogether parties" taking place during the midterm season. The idea was to sign up to host parties for friends or employees or customers on Election Day so people would have a place to celebrate having done their civic duty.

"This is very much part of American electoral tradition," said Green, even if it's something people have forgotten over time. Before reforms in the 1880s sought to protect voters from undue influence and bribery, election days were memorable events. Voters hung out for hours, drinking the free alcohol and enjoying entertainment and the "raucous, freewheeling atmosphere." That raucous atmosphere could go very wrong, though—1800s elections also saw fights, severe voter intimidation, and even riots. Luckily, we've evolved somewhat, and can adopt the idea of the fun without the drunkenness and the danger.

Green and fellow researchers looked at how family-friendly festivals can influence turnout. Experiments involved choosing neighboring towns with similar populations and voting rates and holding a party in one of them, offering cotton candy, music, raffles, and the like. The other towns were left festival-free. In one New Hampshire example, "Election Day Poll Party" was advertised and received press coverage, and perfect spring weather resulted in increased turnout. (Weather matters in a festival's turnout success!) Later replication of the experiment at more locations saw a 2 percent increase in turnout compared to similar cities with no festival.

In the lead-up to the 2016 election, Green worked with the nonprofit Civic Nation to test the theory on even more elaborate festivals—there was more local outreach, and a variety of entertainment at the different sites: dance troupes, photo booths, and even puppies. Looking at nine festivals, including in the battleground states of North Carolina and Ohio, they saw a turnout that was nearly 4 percent higher than in comparable non-party cities. In close elections, that can be a game-changer.

It was this effort that became known as #VoteTogether. Expanded in 2018, it included more than 1,900 parties held in all 50 states. The general idea is for local organizations or chapters of nonpartisan organizations like the YMCA, the United Way, and the NAACP to organize an event tailored to their community. #VoteTogether (which provides a tool kit with all the dos and don'ts) suggests that community parties be held in viewing distance of a polling place on Election Day, and that some type of food, music, and entertainment be provided. In 2018, that meant everything from a few tables with pizza, to a local DJ playing music, to a party with live musicians performing for hundreds.

Shira Miller, who oversees the #VoteTogether effort (it's now under the umbrella of Civic Advisors, which works with When We All Vote), said the long-term goal is for the celebration to become a habit for organizations and for communities. "We are working to change the culture around voting by making it community-driven and celebratory . . . to try to make participation in election day just as celebratory as participating in the Fourth of July," Miller said.

Voter's Illusion Is a Good Thing

It may not be quite as exciting as a party, but every time I get a text from TurboVote reminding me of a registration deadline or a city election, it motivates me to organize my day so I can vote. But it's not just because it's important to support candidates I think are going to do best for my city, state, and country. It's also because of the thought—statistically ridiculous, if still technically plausible—that if I skip voting and everyone else does, too, our democracy would fall apart. If I do the right thing and vote, I tell myself, like-minded people will probably go through the same thought process and show up as well.

This is a form of "voter's illusion." Because elections are rarely decided by a single vote, especially in national elections with millions and millions of voters, we all have to be a bit egocentric to think that our one vote matters. And part of the way we convince ourselves that it does is because we attribute our behavior to others who think like us: if I like Candidate X and take the time to vote for her, others who support Candidate X will do the same.

So what if my vote is just one drop in a very big bucket? It matters. It's me saying who I think should make decisions and what those decisions should be. And if I stay home and you stay home and your dad stays home and so does your best friend and your little sister, then we all lose. I'm voting to do my part for you, and you'll do the same for me. And if embracing our voter's illusion is a delusion of grandeur? That grandeur is what keeps our democracy alive. And earns us a sticker.

VOTING LANGUAGE FACTS

The fashion writer Teri Agins advised colleagues to sprinkle "chocolate chips" throughout their stories—tidbits that people would want to share with others. Below are conversational treats handy to know during election season.

Suffragette—It's often used as a default name for those who fought for women's suffrage, but it's more accurately used only for British women who fought for a woman's right to vote in that country. British journalists first used "suffragette" to make fun of the women and belittle their cause, "-ette" being a suffix the French use to describe something small. Though it originally had a negative connotation, some British women decided to adopt the term. The word was also applied in a derogatory way to American suffragists, but it never gained popularity among women in the United States in the same way.

Ballot—This word also derives from a diminutive—*ballotta* is the Italian word for little ball. In ancient Greece it meant pebbles in an urn; during the Italian Renaissance, people would vote with small balls, often colored for a particular candidate. Casting their *ballotta*, if you will. *Ballotta* became "ballot" in British English. One meaning of "cast" is to throw forcefully, though I've seen no research indicating voters in Venice tried to toss their *ballotta*s particularly hard or from a great distance.

The government-issued printed ballot, and voting in "secret," came along much later. Historian Jill Lepore pegs 1634 as the first time the Massachusetts governor was elected "by papers." But, she adds, "Well after American independence, elections remained widely the stuff of corn and beans and hands and feet." Kentucky voted by voice until 1891.

Party ticket—Even once ballots were on paper, people for a long time brought their own paper to use, or got them from partisans. Pre-printed ballots that looked like train tickets would endorse all candidates from one party—thus "party tickets"—so voters could participate even if they could not read or write. The government-issued ballots as we now know them are an import from Australia, believed to come from an 1856 law and adopted by the British Parliament in 1872. Massachusetts led the "papers" charge again in 1888, becoming the first state in the nation to pass a law requiring the Australian ballot.

What seems reasonable now was a discrimination tool then; the government-issued paper ballot discouraged those who couldn't read from participating.

Chad—Who can think of a paper ballot without thinking of its most infamous failure: in the 2000 presidential election, which left Florida officials holding ballots up to the light, squinting and making controversial calls over whether a hanging "chad" meant a vote for George W. Bush or Al Gore. Merriam-Webster defines chad as "small pieces of paper or cardboard produced in punching paper tape or data cards." Its first usage is listed as 1944, but its origin is unknown.

Other words first used in print that year include the very fitting "gobbledygook" and also "dead presidents," as in cash. (Bonus chocolate chip: during a U.S. Supreme Court argument over partisan gerrymandering in 2017, Chief Justice Roberts called statistical evidence of the impact of gerrymandering "sociological gobbledygook.")

The "I Voted" sticker—*Time* magazine cites 1992 as the likely earliest media mention of the "I Voted" sticker, when the *Miami Herald* reported that Florida businesses were offering discounts to customers wearing theirs. Miami-Dade made sticker news again more than three decades later, with a story about voting stickers getting a "multicultural makeover." It included "I Voted!" in three languages (English, Spanish, Creole), complete with a declarative-statement hashtag: #iamelectionready.

PART THREE

Know Before YOU VOTE

CHAPTER SEVEN

GERRYMANDERING: OVER THE LINE?

Rally to end gerrymandering outside the U.S. Supreme Court, Washington, DC, March 26, 2019 (Courtesy of League of Women Voters)

The only word in the voting lexicon that beats "chad" for absurdity has to be "gerrymander." Its origin story dates to 1812 in Massachusetts, when Governor Elbridge Gerry allowed a redistricting plan that favored his own party and included a state senate district shaped, to a political cartoonist's eye, like a salamander. Gerry + mander = gerrymander. (Gerry name purists, in a GIF/JIF-style argument, insist we should respect Gerry's pronunciation and say "GARYmander.")

It's a complicated word for a complicated problem. Gerrymandering is drawing boundaries of an electoral district to benefit a party or class, and it is one of the hottest topics in politics. Scrutiny from the public and both sides of the political aisle made gerrymandering a focus at two very different high-profile venues in 2019. In March, the U.S. Supreme Court heard arguments over district lines in North Carolina and Maryland. In April, the Tribeca Film Festival featured a documentary detailing political neophyte Katie Fahey's unlikely quest to end partisan gerrymandering in her home state of Michigan.

"Gerrymandering has exploded in the public consciousness in the last two years, particularly on the left, because there was a clear pattern of Republican advantage versus their share of the vote," said David Wasserman, an editor of the nonpartisan The Cook Political Report and an expert in U.S. House of Representatives politics. General public perception tends toward the belief that it's something that's been done for hundreds of years (true), and that politicians of both parties do it when they have the chance (also true). Still, the way it works keeps many voters from understanding exactly what the practice is.

Here's a simple way to think of gerrymandering. Picture a ten-story apartment building that needs to be divided into four "districts," each with an equal number of people. Fifty percent of apartment dwellers prefer Party A, and 50 percent Party B. But Party A gets to decide how to draw the districts. Instead of simply dividing the building into quarters—straight lines, identical shapes—the Party A line-drawer snakes her way around, skipping an apartment here, including one there, based on whether the apartment dwellers prefer Party A or Party B. She's confident she's successfully targeting by party because she has sophisticated mapping technology and demographic info to help her.

After her work is done, the divisions look like four jagged puzzle pieces, but they all contain the same number of people. Still, she's managed to stuff—"pack," in gerrymandering parlance—a ton of Party B people in one district, and spread the rest—"crack" them—throughout the other three "districts." When Election Day arrives, Party A wins three districts by a relatively small but safe margin. Party B wins only one, but by a lot.

It plays out similarly in the real world. For the last several U.S. House elections in North Carolina, the Democrats have won close to half the votes statewide, but only three out of thirteen congressional seats. The lines were drawn in such a way that even with their strong presence in the state, Democrats weren't the majority in most of the congressional districts.

Districts can look quite misshapen in real life, too, zigzagging across a state in nonsensical ways. Some creative types have described recent gerrymandered districts this way: "Wile E. Coyote on a hoverboard behind Peppa Pig" (North Carolina), "Goofy kicking Donald Duck" (Pennsylvania), the "fajita strip" (Texas), and the "broken-winged pterodactyl" (Maryland).

Gerrymandering is an ongoing issue in a handful of states in which the out-of-favor party and anti-gerrymandering advocates felt that the maps drawn over the last decade were blatantly biased. Courts forced some states, among them Pennsylvania, Virginia, and North Carolina, to redraw some of their maps; and states including Michigan, Ohio, Texas, and Wisconsin all saw extended litigation. Still, gerrymandering affects citizens nationwide because it helps to determine who is elected to the U.S. House of Representatives, and thus the balance of its members.

Politicians get their best chance to gerrymander after the U.S.

census, the count of the nation's population that is required every ten years. Census results require the redrawing of districts to correspond to the new population numbers. The stakes go way up in elections during census years, because in many states, voters are selecting officials who will draw lines meant to last a decade. Republican strategist Chris Jankowski said of 2020, which is both a presidential election year and a census year, "It's all the marbles on the table."

Gerrymandering can (and does) happen in drawing the boundaries of the seats for state legislative maps, and it is often in turn the state politicians benefiting from it who then draw the maps for U.S. congressional districts. Much of this chapter will focus on partisan gerrymandering as it relates to U.S. congressional district boundaries. By federal law, those districts must be as close to equal in population as possible; the legal language is as equal as is "practicable." The result is that smaller towns are likely to be grouped with surrounding cities, whereas larger cities may have to be broken up. Some states have additional guidelines, like keeping cities or counties intact when possible.

There are two main kinds of gerrymandering: racial gerrymandering and partisan gerrymandering. The Supreme Court ruled in 1986 that racial gerrymandering is illegal and violates the Voting Rights Act when lines are drawn to lessen the power of a particular group in a way that prevents them from electing a candidate of their choice. That means that an area of predominately black residents shouldn't be split in three ways and grouped with majority white districts for the purpose of diluting the black residents' votes.

"Majority-minority districts"—where a racial minority group makes up the majority of people in the district—comply with federal law in that they prevent dilution of the minority vote and often result in that group's electing a politician of their own race. That scenario sometimes creates unexpected support from conservatives for the creation of such districts. Majority-minority districts usually skew Democratic, which sometimes means more Republican districts surrounding it. In that way, majority-minority districts can mean fewer Democratic districts.

While the causes and effects of racial gerrymandering and partisan gerrymandering are often intertwined, partisan gerrymandering has spent the last several years under scrutiny because the legality of the practice was unsettled and thus the subject of multiple high-profile lawsuits.

Several factors have to come together for politicians to be able to draw lines to favor their own party. First, politicians need to be the ones with the authority to draw the lines. Some states take the line-drawing power completely out of the hands of politicians or require independent oversight. When independent or bipartisan commissions draw the lines, the partisan discord and scheming is usually lessened or even inconsequential. About thirty states require state legislatures to draw their U.S. congressional lines.

In states with a diverse enough political population that a meaningful issue could arise, stars still need to align. All branches of state government will usually need to be governed by the same party in order for that party to dominate the decision-making. If you have a Democratic-controlled state house of representatives and a Republican-controlled state senate, the parties will likely have to compromise to get a map passed. If the same party controls both statehouses but the governor is of a different party, she or he can often veto any unbalanced map. Some exceptions: in a few states, including North Carolina and Connecticut, the governor doesn't have veto power over the drawing of congressional district lines.

Elected officials who have the power to draw lines in their favor still have to predict how people in the new districts would vote. That's the easy part.

In many states, people designate which party they identify with when they register to vote, so sometimes whether you're a Republican or a Democrat is public information. Further, mapmakers can predict voter behavior with near certainty through analysis of census data, demographic information, voting history, and even consumer data. The biggest indicator is a geographic one: urban dwellers are more likely to vote Democratic, while Republican voters dominate rural areas. An oft-referred-to party indicator, made prominent by Wasserman, is whether you live closer to a Whole Foods (likely Dem voters) or to a Cracker Barrel (likely Republican voters). Sophisticated mapping programs compile all the data and offer up thousands of mapping options precise enough to separate a street from its neighborhood or split a college campus.

Why Do We Care Now?

If it's been happening forever and is done by both parties, what makes it a bigger deal now?

Over the last decade, politicians seeking to gerrymander have been particularly good at it, and lawsuits and public attention have grown. A party can win a majority of votes in the state but still win less than half the congressional seats. One state-specific example is Pennsylvania. "Barack Obama carried Pennsylvania in 2008 and 2012 with a roughly similar margin. In 2008, that led to the election of 12 Democratic congressmen. In 2012 it generated five," wrote journalist David Daley in *Ratf**ked*, his fascinating and often shocking 2016 book on partisan gerrymandering.

Daley, also a senior fellow at nonpartisan electoral reform advocacy group FairVote, examined how Republican strategists targeted specific state legislative seats in swing states. Calling their plan "REDMAP," they poured money into 2010 Republican candidates in states including Pennsylvania, North Carolina, Ohio, and Michigan. (Jankowski was one of its primary strategists.) They ran tough campaigns, sometimes using controversial, negative tactics, to win enough seats to give Republicans control of drawing the U.S. congressional maps in those states. It enabled Republicans to secure a large majority in the House in 2012, and they held on to it for the next six years.

The maps drawn in North Carolina have been the subject of years of litigation. Two of the congressional districts were invalidated by the U.S. Supreme Court in 2017 as being unconstitutional racial gerrymanders. The mapmakers denied racial bias, with one of them, the late Dr. Thomas Hofeller, saying during litigation that the goal was "to create as many safe and competitive districts for Republican incumbents or potential candidates as possible." (Hofeller, called the "master of gerrymandering," died in 2018; after his death his computer files became evidence in litigation not just involving gerrymandering, but also over who had advocated to add a question about citizenship to the 2020 U.S. census.)

Wasserman thinks REDMAP's creators get a little too much credit as political masterminds. Republicans enjoyed a wave midterm election year, when 2008 Obama voters didn't turn out in 2010. Democrats had done the same in other wave election years, he said, and the concentration of Democrats in urban areas helped allow Republicans to draw some really favorable maps.

Even without extreme gerrymandering, Wasserman said, it's still difficult to draw maps that perfectly match the Republican/Democratic breakdown of a state. Americans self-select and live near people who tend to vote as they do. Urban centers are likely to be grouped together, while the rest of the districts, spread out across the state in mostly rural areas, are going to be conservative. It's not always easy to draw around that.

Daley disagrees with the position that geography is that determinative, saying that multiple judges at the state and federal level have found maps that were the result of extreme partisan gerrymandering unfair, and accepted evidence of various ways the maps could be drawn to achieve a better representation. He referenced the 2008 Pennsylvania breakdown of twelve Democratic seats to seven Republican ones on maps drawn earlier that decade. Four years later, on maps newly drawn by Republicans and with a similar partisan breakdown in votes, Republicans won thirteen seats to Democrats' five. To assume clustering was the problem, you'd have to believe a lot of people moved to be close to their like-minded friends, Daley said.

So, while experts disagree on the extent to which self-clustering affects line-drawing, there is a difference between that and extreme gerrymandering. And following the 2010 census, Republicans had the power in states that really mattered, including swing states and states inching closer to becoming one. Wasserman gave the rundown: North Carolina, Michigan, Pennsylvania, Ohio, and Texas.

The aggressive gerrymandering in those states meant lots of lawsuits asking courts to rule if the maps were fair or if they needed to be replaced with more representative ones. Judges often ruled that they did need to be replaced. "We had more replacement maps this decade than we had last decade," Wasserman said.

Supreme Court Takes on Partisan Gerrymandering

That brings us to the questions the Supreme Court wrestled with for years: It is clear that some states' district boundaries are drawn to heavily favor one political party in a way that conflicts with citizens' political preferences overall. Is that representative of democracy? Would the Supreme Court put a stop to it? In the summer of 2019, the justices weighed in and answered those two questions: 1) Probably not. 2) No.

Here's the background: After part of North Carolina's U.S. congressional maps drawn after the 2010 census were declared unconstitutional due to racial gerrymandering, North Carolina legislators drew their maps again. They knew they had to solve the racial gerrymandering problem but still wanted to achieve the same representation makeup they had: ten Republican seats and three Democratic seats.

Their plan worked, but they also ended up back in the Supreme Court after they were sued by the League of Women Voters and others, who claimed the maps deprived voters of their democratic right by purposefully diluting their vote. The league argued in its Supreme Court briefs that the maps "cracked and packed" Democratic voters throughout the state, and thus those voters' votes carried less weight than they would in a fairly drawn district.

For its part, North Carolina argued to the court that "the framers assigned the inherently political task of districting to political actors"—state legislatures—and oversight of the fairness to Congress. It's expected it would have partisan influence and results. In short, the argument is that people have a right to vote, but they don't have a right to ensure a win for their candidate.

North Carolina's lawyers also argued that it's too difficult for the court to design a test for how much partisanship is too much. A test for determining what counts as too much partisanship had long eluded the court. When similar partisan gerrymandering cases came to the court in the years before North Carolina's, Justice Anthony Kennedy was repeatedly the swing vote, saying multiple times that there might be a scenario in which he'd rule against partisan gerrymandering, but that he hadn't yet seen a way for the courts to properly oversee it.

In June 2018, Kennedy and his fellow justices punted on ruling whether partisan gerrymandering was constitutional, instead sending back to the lower courts a case over Wisconsin's district lines. Kennedy was expected to be the deciding vote in the North Carolina case, which was the next gerrymandering case on the docket. It was not to be.

Kennedy retired that summer, so when the Supreme Court justices took their seats in March 2019 to hear the North Carolina case, *Rucho v. Common Cause*, Kennedy's replacement, Justice Brett Kavanaugh, turned out to be the one to watch.

It was early spring in Washington, and the cherry trees bloomed pink. The line of tourists hoping to get in to see the arguments wrapped around the building, but the grand plaza in front of the famous Supreme Court steps was empty as I walked to the side entrance designated for press. Reporters were called by seat number and walked quietly and single file, as repeatedly instructed, to chairs at the outer edge of the courtroom.

The seating style—including views obstructed by giant marble columns—and the general seriousness with which everyone took the logistics prior to the proceedings reminded me of covering a fashion show, except instead of spotting movie stars in the front row, it was a mash-up of legal A-listers.

Legendary NPR Supreme Court reporter Nina Totenberg arrived wearing furry brown earmuffs and a colorful scarf. Another longtime court reporter, Joan Biskupic, took her seat just before the arguments started; her newsmaking biography of Chief Justice John Roberts had come out that day. Former attorney general Eric Holder now runs an organization that fights gerrymandering and was also in attendance. There was one movie star: Arnold Schwarzenegger, who took up ending gerrymandering as a cause when he was governor of California.

We scrambled to our feet as the marshal of the Supreme Court announced the justices, calling the traditional "Oyez, oyez" as they took their seats in their tall-back leather chairs. The chief justice sits in the middle and is flanked by the most senior justices. There were moments of levity when Justice Elena Kagan, glasses perched on the tip of her nose, read an opinion about the reach of the National Park Service and announced that an Alaska hunter "could take his hovercraft out of storage." It is, apparently, the hunter's preferred way to cross a river running through a national park.

The court heard two partisan gerrymandering cases that day, the North Carolina dispute first, then a case from Maryland in which Democratic leaders had managed to gerrymander away the one district Republicans usually won in the state. Justice Clarence Thomas remained characteristically silent, but the questions asked by most of the justices gave a preview of how they would vote.

The liberal justices—Kagan, along with Ruth Bader Ginsburg, Sonia Sotomayor, and Stephen Breyer—indicated that they wanted to find a way for the court to stop partisan gerrymandering, with Kagan tempering concerns about a flood of litigation. The reason why "all these politicians are bragging about the amount of partisanship they can put into the maps is because they think it's perfectly legal," she said. "If the Court said it's not legal to do so, presumably, some actors would change their behavior."

It was clear that only two justices' votes were even potentially up for grabs—those of the chief justice and Kavanaugh. Kavanaugh acknowledged that many briefs to the court argued "that extreme partisan gerrymandering is a real problem for our democracy—and I'm not going to dispute that."

But had it come to the point, Kavanaugh asked, that others who could solve the problem, including legislators, state courts, and voters themselves, truly had no likelihood of doing so? Allison Riggs, the attorney arguing for the League of Women Voters, responded with certainty: Yes, the Supreme Court must step in.

Three months later, a majority of the court disagreed. It was a 5–4 decision, with the five conservative justices in the majority and Roberts writing the opinion. "Excessive partisanship in districting leads to results that reasonably seem unjust," he said. But that didn't mean the court could do something about it. "We conclude that partisan gerrymandering claims present political questions beyond the reach of the federal courts."

Justice Kagan wrote a scathing, nearly despairing, dissent. "For the first time in this Nation's history, the majority declares that it can do nothing about an acknowledged constitutional violation because it has searched high and low and cannot find a workable legal standard to apply." She dissented, she wrote, with "deep sadness."

What's Next?

The Supreme Court decision shuts the door on protesting partisanship in maps in federal court, but state court claims will continue. Pennsylvania's gerrymandered maps were thrown out via a state court claim, and "fair map" advocates indicated that they would move the fight to relevant state courts. (State supreme courts can rule on their own state's maps but can't mandate nationwide change.)

The other avenue is in the hands of lawmakers and the people. Multiple times during the Supreme Court arguments, the justices and lawyers brought up efforts by Congress, state legislatures, and citizens to prevent and end partisan gerrymandering via ballot initiatives or state constitutional amendments. Proposals differ from state to state, but the general idea is to remove or at least lessen the partisanship angle by having a bipartisan commission, or a commission made up of Republicans, Democrats, and Independents, draw the lines.

When Schwarzenegger was governor of California, for example, he felt that California's lines were drawn to protect incumbents. He supported ballot initiatives that, when they passed, took the power to draw district lines from state legislators and gave it to a bipartisan citizen redistricting commission.

The attorneys arguing the North Carolina case pointed out that ballot initiatives are not an option in states where laws and procedures don't allow for them. As Riggs said sharply to a combative Justice Neil Gorsuch, North Carolina had no such option. "This isn't self-correcting," she said.

But in states that do allow it and have put it on the ballot, voters have embraced the option. In 2018, Missouri, Colorado, Utah, and Michigan voters approved switching from legislative-drawn lines to commission-drawn ones. (Utah's ballot initiative passed by only 1 percent, but the other states approved them more decisively.)

That's not necessarily surprising, considering how people feel when asked about line-drawing that puts the power to draw political maps in the hands of politicians. In a bipartisan poll of likely 2020 voters commissioned by the Campaign Legal Center, 65 percent said they support removing partisan bias from congressional redistricting, even if it means their preferred political party will win fewer seats.

The Campaign Legal Center's head litigator is legendary Supreme Court specialist Paul Smith, who's argued more than twenty cases in front of the court on multiple topics, including the First Amendment and voter ID; his most high-profile win is the landmark gay rights case *Lawrence v. Texas*. Smith served as counsel of record for the League of Women Voters in *Rucho*.

Standing on the Supreme Court plaza after the arguments, Smith said he was "cautiously optimistic" the court would find a way to curb partisan gerrymandering. But when they didn't, he and other anti-gerrymandering advocates immediately began exploring the state court and legislative avenues, and all eyes turned to litigation already ongoing.

Within weeks of the Supreme Court decision, a trial began in North Carolina state court over legislative maps. A unanimous three-judge panel found in *Common Cause v. Lewis* that the extreme partisan gerrymandering of legislative districts violated the state constitution, which requires that "all elections be free" and which the court interpreted to mean that elections should represent the will of the people.

For anti-gerrymandering advocates, *Rucho* was certainly a setback, but it and the cases before it also pushed the issue into the public consciousness. "I do think the litigation has sparked the popular uprising," Smith said.

A Twenty-Seven-Year-Old Slays the Gerrymander Dragon

Katie Fahey was working for the Michigan Recycling Coalition and feeling disheartened after the contentiousness of the 2016 election. Fahey, then twenty-seven, believed gerrymandering was partly to blame for some of the divisiveness in her state, and she took her frustration to Facebook. She wanted to take on gerrymandering in Michigan, she wrote. Did anyone want to help? She added the smiley-face emoji and posted, and got a lot of replies.

Fahey didn't necessarily know at the time how to stop gerrymandering, but she figured that there had to be a way. In Michigan, it was done by ballot initiative, where voters could approve a change to the state constitution to mandate a commission with a mix of Republicans, Democrats, and Independents to draw Michigan's congressional map. She needed 315,000 signatures from Michigan citizens to get it on the ballot, so she and her army of volunteers held 33 town halls in 33 days to educate people all over the state and hear what those who attended wanted in an amendment.

She and her volunteers chose the name "Voters Not Politicians" and wore t-shirts reading "Slay the Dragon," with a drawing of the nineteenth-century Massachusetts district from which gerrymandering got its name. The team went to carnivals, highway rest stops, churches, and parades armed with clipboards with the outline of whatever district they sought signatures for on the back.

If someone had a question about what gerrymandering meant in practice, Fahey said, volunteers could flip the clipboard over and ask, "Do you think this is drawn in a way to represent your community?"

They delivered 425,000 signatures before the deadline and fought a lawsuit backed by the Michigan Chamber of Commerce that claimed they couldn't be on the ballot at all. That went all the way to the state supreme court, and they won. Fahey regularly updated fans of the movement via Facebook video—expressing jubilation after achieving a certain number of signatures, shedding tears during the litigation when it looked like all their hard work might be tossed aside.

Election night arrived after months of traversing the state, and little sleep. When the polls closed, Fahey and five thousand of her now closest friends had nothing to do but wait. They hosted a viewing party in Detroit but also "had gerrymandered the state into fourteen different regions," she said, with watch parties in each. In Detroit, Fahey paced nervously until a young woman approached, telling her they had won with more than 60 percent of the votes and a winning spread of about a million voters.

"It felt amazing, especially because of how we had done it," Fahey said. "So many people had said that regular people are never going to care about this issue and it's not going to pass." But sadness quickly set in when she thought about how much work it had taken to change something that seemed to her inherently unfair and that so many people were against. Plus, she said, there was "the fact that we've just lived with this since the founding of our nation."

Because it's politics, the battle in Michigan isn't over. A lawsuit claims the commission is unconstitutional, and Voters Not Politicians is now fighting those claims and working to make sure the commission is properly funded, with the goal for it to be active by 2022.

Fahey's journey was the heart of the gerrymandering documentary *Slay the Dragon*, which premiered in Manhattan at the Tribeca Film Festival. She arrived wearing her uniform of a neat blazer and a Slay the Dragon t-shirt and posed on the red carpet with the filmmakers, Barak Goodman and Chris Durrance. The film also highlights the backroom deal-making that led to partisan line-drawing in places like Wisconsin and North Carolina, and the repeated setbacks those fighting gerrymandering faced in the Supreme Court. *Variety* called it the most important political film of the year and said "it may prove to be one of the key political films of the decade."

Making a complicated topic like gerrymandering accessible to a mainstream audience wasn't as difficult as one might think, the filmmakers said, because the conflict feels less like a fight over complicated population numbers and maps and more about the interests of the citizens versus elected officials. It troubled them that because of the way gerrymandering usually happens, out of the public's sight and well before elections, it wasn't something most citizens even knew might affect them. "It wasn't part of what people thought the debate was—it wasn't part of elections; it wasn't part of election campaigns," Durrance said. "It made us really want to figure out ways of telling the story to bring it to life."

Slay the Dragon was acquired by Magnolia Pictures, one of the most well-known independent film distributors and the team that released the Oscar-nominated documentary *RBG*, about Justice Ginsburg. The film was selected for multiple festivals, including in Indiana, Pennsylvania, Michigan, and Arizona. Magnolia planned to release the film in top markets and through video-on-demand in the spring of 2020.

Duly taking note of audiences' overall political fatigue, Magnolia had planned to hold off on putting out any political films. But executives made an exception after being won over by *Slay the Dragon*'s optimism. It provides "a rallying cry about voting equality, and it really ends with this message of 'you can make the change,'" said Magnolia's Neal Block.

The Tribeca premier audience, a mix of film buffs, legal experts, opinion journalists, and Fahey's fellow volunteers, made clear their desire for change. As the credits rolled, a single yelled comment from the back of the theater caused the audience to burst into applause. "Katie for President!"

To 2020 . . .

Gone are the days of gerrymandering as an obscure issue. Organizations fighting gerrymandering are hoping it's one that voters consider as they go to the polls. Partisan groups like Stacey Abrams's Fair Fight and All On The Line, supported by Barack Obama and Eric Holder, made formal efforts to train Democratic voters in how to fight gerrymandering. The League of Women Voters also launched a nonpartisan program called People Powered Fair Maps to instruct voters on ways to curb gerrymandering in their state, including legislative fixes, ballot initiatives, and civic engagement.

I asked Republican strategist Jankowski what'd he say to people who believe it's fundamentally unfair when politicians draw lines designed to protect their party and themselves. He reiterated the arguments made by North Carolina in the Supreme Court—that it's the way the framers set up our system, that they knew decisions made by politicians would naturally be political, and that citizens can work to change the laws in their state. But for those who have a problem with it, he said, there's a solution.

"Vote. Period. Every concern or issue you have about the system can be impacted by a shift in voting behavior, [with] increased voting behavior." What that means is that if turnout increases significantly, lines drawn to benefit one party may not hold. Lines are drawn by predicting who will vote, and if young people come out in droves and other new voters turn out, they can perhaps pull off a surprise. "We thought we'd built a ten-year majority with those lines. But you just never know," he said.

Critics of that argument say some of the gerrymandered maps, like the ones the courts threw out in North Carolina, can withstand any "wave" of voters. When gerrymandering is done effectively, it takes a lot to combat it. What gerrymandering undoubtedly shows is that every race matters. The U.S. House of Representatives supported a law in 2018 that would curb partisan gerrymandering, but it didn't have enough support in the U.S. Senate. And those down-ballot races you usually hear about only in your local news? They may decide who draws the districts in your state, and the officials from your state affect the whole country. Decide where you stand, and vote.

CHAPTER EIGHT

KNOWING THE NEWS IS REAL

(Sean Gladwell/Getty Images)

In August 1765, an angry mob axed their way into and looted—down to the wainscot and children's clothes—one of Boston's finest homes, a multistory mansion belonging to Lieutenant Governor Thomas Hutchinson, a British loyalist. Hutchinson was a regular target of partisan newspapers that criticized Britain's rule over the colonies. The papers exaggerated his position and his views, and the attack was carried out to protest a new tax on colonists. As it turned out, Hutchinson didn't support the tax. The destruction of his home is a colonial anecdote illustrating the consequences of "fake news."

Having to decipher what can be trusted in political news is a problem older than the country, and the stakes have always been high. But certainly early Americans didn't have to navigate the barrage of breaking news, pundits' hot takes, and stories from who-knows-where shared by friends and family in their social media feeds. As elections near, the frequency of new information only increases. Understanding the news today, voting activist and actress Yara Shahidi said, requires deciphering so much political jargon that it can act as a barrier to being socially engaged.

In the last presidential cycle, 91 percent of Americans said that they had learned something about the election in the week prior. That was in January 2016, more than nine months before Election Day. About the same percentage said that the most helpful information came from sources such as cable news, newspapers both online and in print, and social media. Of course, being blasted with all that content takes a toll: with more than 100 days to go before that election, nearly 60 percent of Americans reported being "exhausted by the amount of election coverage."

There is good news. First, it is by no means a requirement to be a news junkie. You don't need to mainline Twitter to be an educated voter who makes informed choices about candidates and issues. Second, there are plenty of sources you can trust, and tactics that can help separate the news from the noise.

This chapter isn't an explanation of how Russian used trolls to do its bidding on social media or step-by-step directions to spot "deep fake" videos that purport to be something they're not, though certainly there's no shortage of altered footage of politicians on Facebook. Instead, the point here is to help you access the news and information you need to make informed voting decisions—ideally without wanting to hurl your phone off a bridge before Election Day.

Generally speaking, the more politically aware you are and the more you trust the mainstream media, according to research, the better you are at telling fact from opinion. Americans who had "high political awareness" and placed "high levels of trust in the news media" were much more likely to be able to distinguish factual news statements from opinion, a Pew study found. (Digital savvy was a big indicator, too.) But even people possessing those qualities had room to learn; less than forty percent of those who have "high levels of political awareness" or "a lot of trust in the information from national news organizations" were able to correctly identify five out of five factual news or opinion statements. Factual statements included "Health care costs per person in the U.S. are highest in the developed world," and opinion statements included "Abortion should be legal in most cases."

"We have more incredible information literally available at our fingertips than ever before, but it's competing with so much more information intended to sell, persuade, mislead, incite, and misinform," said Alan Miller, award-winning journalist and founder of the News Literacy Project, a nonprofit organization that helps educators give students the tools to become informed and civic-minded news consumers.

CNN chief media correspondent Brian Stelter put the challenges news consumers face in a more colorful way. "We're all walking into this overwhelming, intoxicating casino, every day . . . It's open at all hours. There are all these flashing lights, all of these stimuli that are encouraging us to stay."

What's on Your Ballot?

The first step to both becoming educated and minimizing the political news insanity is to focus: find out which elections are on the horizon and what will be on your local ballot. Your local county board of elections website should have an annual calendar for all elections, including local, state, and national, but additional information is available on your secretary of state's website. (If you need help finding your relevant sites, USA.gov is a useful resource.) Local Republican and Democratic Party websites are likely to have similar calendars. Sites like the League of Women Voters' VOTE411 and Ballotpedia often have this information as well. All of these sites help you know which candidates and issues you should follow: Is your state legislator up for reelection this year? Who are your U.S. congressional candidates? Does your city have a mayoral race?

The earlier you check that out (try now!), the more time you have to keep an eye on candidates' positions and the issues likely to be factors in the race. Once you know which races to follow, start taking in the relevant news. But where? And how much? As we all know, it's complicated, even for the most diligent news followers and journalists. While news about major races is likely to come to you via any visit to a news site's homepage or social media, for lesser-known races—city council, school board, judges—even the most active reader is likely to have to seek out detail. But no matter how you come across the information, you have to make a judgment about its value. Consider what you need news sources to provide you with, and the standards you need them to meet.

The Role of a Free Press

The founders of the country believed the role of the press to be vital enough to merit mention in the Bill of Rights. (That doesn't mean that they didn't complain about coverage of them at times; they did. They just knew it was integral to democracy.) The First Amendment says that Congress cannot pass laws "abridging" the press. That's been interpreted broadly over time to mean that the government cannot interfere with the press doing its job, except in extraordinary circumstances. It can't jail a reporter for publishing a critical story, or shut down a newspaper because a politician doesn't like what it wrote.

The press works for the public. It's not unlike how government officials serve the public. The job of the press is to inform you of things going on in your community, your city, your country, and the world. Underlying it all is a commitment to telling the truth and, in the realm of government, a responsibility to act as a check on those in office and those seeking office.

If a candidate is taking bribes or city officials are hiding knowledge of toxic chemicals seeping into the groundwater, the press should help hold them accountable by telling the public about it. Political coverage also serves to highlight the details people want to know about candidates, or, more accurately, all the different things all the different people want to know. That includes how candidates think about issues, what experts and those who will be affected think of candidates' plans, and how those ideas compare to opponents' platforms. Reporters also want to draw for voters a full picture of the candidates—their childhood experiences, their hobbies, how they react under pressure, early jobs that may have informed their opinions.

Sources that provide information supportive of candidates include the candidates' own websites and speeches, partisan commentary, and material put out by activist organizations. All of this should be part of your political content consumption. But it's also imperative to read unbiased news sources. They'll give you the full picture: what the candidates want you to know, and sometimes what they wish you didn't. And they'll allow you to evaluate what you believe—without telling you how to feel. A steady diet of news that gives it to you straight is a voter's superpower.

News organizations are not perfect. Reporters sometimes make mistakes on facts and bad judgment calls on what should and shouldn't be covered, and may frame things in ways that are regrettable after a story fully unfolds. But that's not a reason to mistrust all news coverage. Read with a critical eye, but not with one so cynical that you miss learning important things about candidates and elections. A good news organization always corrects its mistakes, quickly and prominently (and should be called on it when it doesn't).

VOTING TIP: Headlines are not the news. Headlines are like a preview of a television drama. It gives you an idea of what the story is about, but it's no substitute for watching the whole thing. If you've only read the headline, you don't know the story.

NEWS YOU CAN TRUST?

As you move through the casino of politics-related information, figure out which outlets you can feel comfortable trusting. And then go to those outlets first to get the facts. Remember that traditional print sources almost all have very strong presences online, and that anything that's in print can usually also be found online.

A place to start in identifying those trustworthy stalwarts is to consider an outlet's history. "A track record counts for something. There's a reason why the Associated Press and the *New York Times* and NBC and ABC and Reuters and other big, old-fashioned news outlets remain among the most trusted. And that's because they've been doing it a long time . . . and they have a reputation for trying to get it right," CNN's Stelter said.

The same goes for regional and big-city papers: the *Houston Chronicle* (founded in 1901), the *Omaha World-Herald* (1885), the *Salt Lake Tribune* (1870), and the *Sacramento Bee* (1857). Small-town papers and local radio stations often have similar legacies, though most papers are having to do so much more with many fewer subscribers. Local television stations also strive to bring you accurate information, though a caveat in today's media landscape is that their national ownership may direct coverage and commentary more than is obvious from watching. The three major television networks—ABC, NBC, and CBS—are probably the most down-the-middle of national television coverage, and their morning and evening news programs hit the highlights of national stories and high-profile state races. (CNN is the most neutral of the cable channels.)

Places you can count on for the nitty-gritty, nearly hour-to-hour fact updates on national races? The newswires, including Reuters and the Associated Press. Those organizations send reporters to cover presidential candidates essentially anytime they're in public, and have reporters covering the U.S. House and Senate full-time as well. If a candidate said something publicly or did anything new or unusual on the campaign trail, those services are very likely to have the account or will very quickly confirm it. Politics-specific sites like Politico also are well sourced with candidates and focus on wall-to-wall political coverage. FactCheck.org is a great place to go for detailed fact-or-fiction breakdowns on controversial statements.

There are of course many magazines with track records for accuracy—big-name ones like *The New Yorker,* with its rigorous fact-checking team, and *Time*, which is still alive and kicking almost 100 years in. Some state citizens are lucky enough to have magazines dedicated to them—*Texas Monthly*, for instance, has kept Texas politicians honest (and called them Bum Steers when they weren't) for decades.

Especially in election years, countless books focus on both candidates and newsworthy policy issues: education policy, immigration, climate change, gun control. The best way to evaluate a book is to check the bio of the author—what's their background, how do they know what they know, and do they have skin in the game that might make them want to convince you one way or the other? Reading reviews of the book will also tell you how it was received—did other experts in the field find it credible? Did it ignore or gloss over a flaw in its premise? (As with other news types, the answers to those questions don't necessarily mean you should or shouldn't read, just that you should read with an understanding of whether the information is coming to you with an angle.)

Podcasts have blown up over the last few years and are a great way to listen to news from traditional outlets, news radio like National Public Radio, and experts and commentators on nearly any topic imaginable. Check the hosts' bios and their other work to evaluate both their trustworthiness and any biases.

Focusing on household-name news organizations leaves out countless news sites that provide straightforward, accurate, informative political coverage. But—just as with the bigger ones—it'd be impossible to list them all. If you know how to evaluate a news story—

whether it comes to you online, in print, via video, or in your earbuds—you'll be able to discern what you can trust.

The Hallmarks of a Proper News Story

News stories—not opinion, not commentary, but straight-up news stories—answer the who, what, where, when, and why questions. They serve to inform, not to convince. They might cover breaking news (who won the New Hampshire primary, or a governor resigning under a cloud of scandal). They might be feature stories (how candidates' nonpolitical employment history affects their policy ideas) or investigative pieces that have taken months and months to report. They try to present facts, explain the different sides of an argument, and put the information into the context of the larger political climate or race.

When you read a news story, questions to ask include: Where is this person getting the information? Did he or she see it firsthand? Was the reporter at the press conference or congressional hearing, or did he or she talk to the people who did attend? If you're reading online and the story is a rehash of an article that links to the original, be sure to check out that source as well.

Michael Schmidt has been a *New York Times* reporter for more than a decade and covers national security and federal investigations. Schmidt was part of two reporting teams that won Pulitzers, the American journalism world's top prize, in 2018—one for coverage of the Trump administration and the other for stories about sexual harassment.

Schmidt took an hour's break from reporting to discuss with me how he builds stories. We settled into the bland bustle of Le Pain

Quotidien near the *Times*'s Washington bureau for our interview. Like all reporters working today, he's keenly aware of distrust in media, though even that involves deciphering what "media" means. Newspapers? Partisans on cable news shows? Anyone with a Twitter account? "We all get lumped together, and we all don't have the same standards, biases, or beliefs," he said.

There is no official list of rules all media follow, but there are codes of ethics and basic tenets that guide any reputable journalist. The simplest is don't put information in the paper or online or on television or on a podcast unless you're as positive as you can possibly be that it's true. But also make sure that you've sought comment from the parties involved, that you've included all relevant information, and that you've made clear what you know and what you don't.

"We [need] to do more explaining of what we actually do," Schmidt said, to help the public understand "that we're nonpartisan fact-finders that go out and try and take a snapshot of the world every day and tell you what's going on. It's an imperfect medium. It's a very difficult thing to make a judgment about something on a day-to-day basis. But it is not a partisan exercise. It's an exercise in trying to dig as far and hard into the facts as possible."

So how does a reporter do this? In the comic-book version, Clark Kent sits at his desk at the *Daily Planet* waiting for a scoop. No disrespect to Superman, but that is definitely not how it works. Political reporters covering a candidate or race work to get to know not just the candidate but the campaign staff, volunteers, people the candidate used to work with, election officials, opponents' staff, subject matter experts who can provide insight on policy, and on and on.

Building a network of sources and learning about your subject is the only way a reporter can be successful at informing the public, Schmidt said. "We're not omnipotent . . . We have to use the tools that we have. And that is to go out and see things, to listen to things, and to build relationships with people that will provide us with a greater understanding of something."

Another truth about organizations with a big reach and a long history is that reporters on staff often have an easier time getting sources to return their calls, texts, and emails. When requests for an interview or information come from the *Wall Street Journal*, the *Washington Post*, or CBS News, sources are likely to respond. This means that well-established publications often have greater access to the people in the room, which often allows more in-depth coverage. (There are obviously exceptions to every rule, and sometimes sources prefer to avoid the tough questions, or leave out something they'd rather not admit. But a good reporter works around that to develop more sources and find more people who know what happened.) For local stories, reporters from newspapers, news sites, radio, or television stations from the area almost always have the best information—familiarity with the landscape and access to sources are keys to a good news story.

The bottom line is this: Don't use pundits as your primary news source. When you're reading or watching or listening to political news, the stories you can trust the most let you know that the news organization delivering the information had reporters gathering that information—they witnessed the event or talked to people who did, talked to people who disagreed, and can give you context as to why it mattered or didn't. You want your news to come as straight from the reporter to you as possible.

Spotting an Unfair Story, and What Happens When a Reporter Gets It Wrong

Before a story is published, reputable news organizations give people who are the story's focus or are mentioned in a way they might take issue with a chance to respond. "No surprises!" is a common refrain among editors. Subjects of unflattering coverage should not read the news and be shocked by what they see about themselves.

Reporters often have the tough task of telling people that a story is going to run that is critical of the way they do their job. If you're reading a story that seems particularly harsh or critical, always look to see how the subject responded. If you don't see quotes from subjects or their organizations, or a sentence that they declined to comment or couldn't be reached, or, at the very least, some representation of their opinion or position on the matter at hand, it's a sign that the reporter didn't do his or her job and that the news organization isn't one that should be trusted as a primary news source.

Despite all the checks in place, mistakes happen. It can be as simple as a misspelled name or incorrect title, or something substantive, like incorrectly describing a politician's position on a controversial issue.

"Reporters live in fear of making mistakes. There's nothing more embarrassing," said Stelter. Schmidt echoed to me just how awful it feels to get something wrong, but also why running a correction as soon as possible is important. "We correct our mistakes. And it's a very painful thing for us, because we're basically admitting that we failed to do our job," Schmidt said. But admitting errors shows "the devotion that we have to trying to follow the facts and get it right."

If you read something that doesn't meet all the standards reputable news organizations follow, see if another news organization has the same story. In the fast-moving world of political news, if one outlet airs or publishes a newsworthy item about a national, statewide, or big-city race, other outlets will seek to independently confirm the story and "match" it as soon as possible. It's usually within hours, and often within minutes. Rarely does it take more than a day. If the story involves a smaller town, the news may come only from that town's main news source. But if it's a true story, including one that starts from local eyewitnesses sharing information on social media, the story is likely to be built upon by the news organization. Watch for follow-up stories, how the relevant campaign responded, and what the opposing side says as well.

Diversify, Diversify

Most people's media diet consists at least in part of outlets or reporters that tend to support their points of view, including, of course, cable news. Americans are increasingly likely to gravitate to news and information that aligns with the political positions they already have, both by selecting it on their own and by social media services' algorithms prioritizing it in their feed. It's important to diversify the news you read, watch, and listen to, and to be aware that hyperpartisan rhetoric can be misleading.

In the lead-up to the 2016 election, a *BuzzFeed* analysis found that hyperpartisan Facebook pages regularly trafficked in false information, and that false information has legs:

> *The review of more than 1,000 posts from six large hyperpartisan Facebook pages selected from the right and from the left also found that the least accurate pages generated some of the highest numbers of shares, reactions, and comments on Facebook—far more than the three large mainstream political news pages analyzed for comparison.*
>
> *Our analysis of three hyperpartisan right-wing Facebook pages found that 38% of all posts were either a mixture of true and false or mostly false, compared to 19% of posts from three hyperpartisan left-wing pages that were either a mixture of true and false or mostly false.*

I'm not saying to never look at sites or channels that report their news in a way that conforms to your views—following people whose opinions you trust is part of forming an understanding of issues. The key is making that information only a slice of your news intake. Take in information you agree with, information that's down the middle, and information you disagree with, too.

My quick take on when to totally discard what someone's saying is this: If they're yelling at you, acting as if there's only one way to think about a story or issue, or portraying someone as a hero with no faults, they're doing you a disservice. If you see a "news" organization doing that—change the channel, close the browser, or log off.

HOW TO WATCH A CABLE NEWS PANEL

Turning on cable news often means seeing a panel of multiple people stretched across a long table with laptops in front of them, or maybe multiple people with their heads in individual boxes, fake skylines behind them. If you're like me, your first thought is, "Who are all these people?" It is almost impossible to figure out who's who, or what roles they are playing in analyzing and commenting on the news. Here is a simple breakdown of what various titles mean:

Correspondent: The reporter and fact-gatherer. This is usually the person who interviewed the source or attended the event being discussed. The correspondent's job is to relay the news in as straight a manner as possible. Correspondents provide context on why a story matters or events that led to it, but they don't usually share their opinions.

Anchor: The person may not be identified that way, but on television this is the person leading the coverage. For CNN, that could be Anderson Cooper or Jake Tapper. Anchors might report as well, but are there to ask questions of guests and steer the ship. They provide analysis and context. Though they also usually shy away from sharing their personal opinions, they ask the questions that people at home might have, or play devil's advocate to explore all sides of a story.

Analyst: These are people with subject-matter expertise who provide context for the news. CNN legal analyst and *New Yorker* writer Jeffrey Toobin, for instance, provides historical precedent and analysis of Supreme Court opinions. Analysts are likely to offer their opinion—they might say what they think will happen next or how people are likely to respond, but they aren't affiliated with any particular player or party.

Contributor or Commentator: These are people who are being paid to provide opinions, and they'll often be affiliated with an outside group—a commentator who works with or speaks in service of the Republican or Democratic Party, for example.

Cable news should identify the backgrounds or biases of particular commentators more frequently than they do, but if someone is talking and you can't figure out their angle, use the "second screen" that's likely at your side—your laptop or another device—and google the speaker to check their credentials.

Non-News Sources

In political races, it's always important to spend some time on the candidates' own websites. It helps you know what they prioritize and what they see as their strengths. Candidates' websites usually lay out their positions on prominent issues and the plans they hope to enact if elected. It's also the place to learn if candidates are hosting meet-and-greets or town halls where you can hear their views and possibly ask questions.

The League of Women Voters' VOTE411 is a site that helps distinguish between candidates. It includes questionnaires, with candidates for a particular office all asked the same questions, including on topics specific to their region. Jeanette Senecal, the league's senior director of mission impact, highlights ballot initiatives as deserving extra attention before you vote. Ballot initiatives ask voters to approve or deny a change to law or an action the area in which you live is proposing to take. They cover a wide array of issues, among them voting rights, city development, and campaign finance. Gay marriage was once a ballot initiative issue in California. In Florida, voters were asked if felons should be allowed to vote. A ballot initiative in North Dakota asked if volunteer fire fighters should have special license plates.

It's necessary to find clear language about both what the ballot initiative is asking and what a yes or no vote means. Ballot initiatives are often worded ambiguously, in a way that encourages people to vote the way the author of the initiative wants them to. Confusion over language can lead to voters selecting the opposite of what they really want, Senecal said. VOTE411 provides clear descriptions and explanations for statewide and many local initiatives, as do Ballotpedia and some secretary of state websites.

Senecal also advises watching candidate debates, and she has suggestions for how to watch in a thoughtful way. Presidential-candidate debates obviously get the most attention, but debates are held for many other offices, too—Congress, state legislature, and mayoral races included. It's a good opportunity to add some fun to the process by asking friends to join you to watch, whether you're tuning in on TV or streaming. Consider in advance of the debate which issues you most care about so you're listening closely for comments and questions related to them. When the debate ends, turn the TV off for a few minutes to take some time to decide for yourself how you felt it went, without pundits weighing in. Who made better points? Which side is most aligned with your values and needs? If you have friends with you, discuss it with them.

Well-informed friends can be helpful in preparing you for the polls. Patagonia's Corley Kenna hosted a "ballot party" to go over the notoriously long California ballot. (Feeling informed about certain races took some work for even the very plugged-in Kenna.) A friend of Clique's Hillary Kerr sends scripts for what to say to congresspeople you want to call about a particular issue. Many people rely on trusted friends or family or colleagues who take the time to break down local ballots and are willing to share the info. Ask your most politically engaged and trustworthy friends how they prepare. And if they don't already write it all down, offer to work with them to do that, and to share it with friends.

Another way to decipher tough issues is to research sites and organizations that focus on that topic. If you find an organization you trust, following them can be a great way to help crystallize your thinking on various issues. For judicial elections, for example, state or local bar associations might offer nonpartisan information on candidates, and in some cases might provide lists of candidates' qualifications and previous judicial experience. Many issues sites will take a position, so just be sure to confirm and understand who's funding the site. "Who is trying to influence your perception of a candidate or that issue? It's important to know who's behind the information that's in front of you," Senecal said.

Again, that's the central question for any information you come across when preparing to vote, news or otherwise. At a news literacy conference I attended, a teacher said she's constantly telling her students, "Google is not a source." True! The same goes for Twitter, Facebook, and a friend summarizing something they read. The source is the original news report or campaign statement or activist group's research. Find that source, and ask yourself if it comes with any biases. Being an educated voter is something to take seriously. And, like getting a second opinion on a medical diagnosis, when it comes to information about candidates and issues, it's always good to check a second source.

CHAPTER NINE

UNDERSTANDING POLLING

Hillary Clinton supporters watch election returns, New York, New York, November 8, 2016 (Angela Weiss/Getty Images)

I met two friends for a pizza lunch in Brooklyn on Election Day 2016, and we spent our time chatting about how the country was most likely twelve hours away from electing a woman as president. Expert after expert was saying that Hillary Clinton's chances of winning were over 80 percent, and women were visiting the grave of Susan B. Anthony, leaving their "I Voted" stickers on her headstone. On the subway ride home, I listened to FiveThirtyEight's end-of-campaign podcast, and a few of the hosts—whip-smart data lovers all—seemed confident Clinton would win. But the site's editor-in-chief, Nate Silver, who in 2012 correctly predicted which presidential candidate would win each state, was hedging.

That skepticism stuck with me as that 80 percent chance of a Clinton win began looking questionable early in the night; by early morning, the chance was zero. And it took zero seconds for people to start asking how the polls got it so very, very wrong.

While pollsters and analysts certainly couldn't say they got it exactly right, they collectively spent much of the immediate election aftermath saying they weren't so wrong, either. They tried to drive home that a close race is a close race, and that an 80 percent chance of winning still means a very real chance of losing.

It's tempting to think that we should all just ignore the polls and move on with our lives. But polls have a circular effect on politics—politicians depend on outside polls and run their own internal ones to help shape their policies, strategy, and messaging. News organizations also conduct their own polls and tell us about them, and about others, because the polls provide insight into how Americans are feeling and indicate how campaigns might shape their policies, strategies, and messaging.

Polls also have a huge effect on how campaigns spend their money. Asking how voters respond to particular scenarios, like a candidate's experience as a prosecutor, can tell a campaign if the information makes voters more likely to vote for the candidate. Campaigns don't want to spend money on expensive TV ads unless they know that the information in the ad will help the candidate, said Philip Bump, a national correspondent for the *Washington Post* who focuses on the data behind politics.

If you're interested in election news at all, polls are nearly impossible to avoid. Plus, you're likely already interested and have been for quite a while. A Fox News poll all the way back in April 2019 found that 52 percent of Americans classified themselves as "extremely" interested in the upcoming presidential election. That's a level usually not achieved until the weeks before an election. If we simply won't avoid them, we owe it to ourselves to make sure we know which polls to trust and how to give them proper context.

A first question many have is how a poll can represent a whole group of people if only a few are asked the questions. The answer is, it's an art and a science. Reputable pollsters use complicated methodology to create a composite of the group they're measuring by selecting people based on their different characteristics—party affiliation, gender, geographic location, age, education level, and countless other variables—and then give proper weight to how represented that group is in the broader electorate. You can't call two hundred people in largely liberal San Francisco or mostly conservative rural Alabama and expect it to represent the entire country. But if you have the correct mix, a small sample can tell you something about the broader world. A larger sample size can tell you a bit more. Polling experts like to use this analogy: a doctor doesn't need to test all your blood to find out if something is wrong with you. Just a sample will suffice.

After the 2016 election, the polling world did a significant post-mortem to figure out what they got right and wrong, and what in the world happened in general. A comprehensive evaluation was done by the American Association for Public Opinion Research, with analysis and input from top professors, journalists, professional polling organizations, and others who evaluate election-related polls.

2016: What Went Wrong?

What they found was that the national polls—the ones that looked at how the country would vote overall—were actually quite accurate and historically in line with how national polls performed in prior presidential elections. "Collectively, they indicated that Clinton had about a 3 percentage point lead, and they were basically correct; she ultimately won the popular vote by 2 percentage points," the report said.

State polls, however, had a historically bad year, with an average absolute error rate of 5.1 percent, the largest in presidential elections starting from 2000. The fact that there are fewer local news organizations meant there were fewer people paying for objective polls, and many of the state polls that were conducted weren't well funded. States that were competitive did actually have a lower error rate than noncompetitive states. Even so, state polls said Clinton had a narrow lead in Michigan, Pennsylvania, and Wisconsin, but Trump ended up winning all three. Those three states were determinative as to who won the Electoral College.

A few major factors caused the polls to be off. First, people changed or made up their minds close to Election Day. "There is evidence of real late change in voter preferences in Trump's favor in the last week or so of the campaign, especially in the states where Trump won narrowly," the report said. Pollsters develop the numbers on "inconsistent responders"—those who changed their minds—by calling back the same people they called before the election and asking if they voted how they'd said they would.

In most elections, voters who change their minds often "wash out," the report said, breaking even between those who switch from Democrat to Republican and vice versa. But in 2016, these inconsistent responders broke for Trump by a 16-point margin. That is more than double the second-largest margin, a seven-point voter break for George W. Bush in 2000.

Another main reason for poll miscalculations is that voters' levels of education turned out to be a huge predictor of voting preference in this election. (In general, people with a higher level of education voted for Clinton.) Unfortunately, some pollsters didn't properly adjust their methods of evaluation for that factor, even though college graduates were overrepresented in their samplings.

The analysis points out that election polling is always difficult. Not only do pollsters have to find a representative mix of people, but they have to attempt to predict who will actually vote. Far out from an election, polls might start with eligible voters (generally Americans eighteen and over), then narrow to registered voters as an election gets closer, and then again to likely voters. Because people say they'll vote and then don't make it to the polls, even that last category isn't a sure thing. It's part of the reason pollsters say they're giving you a snapshot in time rather than predicting the future.

Leading up to the 2018 midterms, Katie Couric interviewed FiveThirtyEight senior political reporter Clare Malone and asked her to explain what went wrong in 2016. (A "big whiff," Couric called it.) Malone said that FiveThirtyEight's calculation that Clinton had a 71 percent chance of winning made them the least wrong. But, she said, "we were, and all of the other models were, off." And since that election, she said, she and others in the data reporting world are hoping to educate people to become more "literate readers of polls and to know we're not pulling this out of nowhere. And we also want you to know that there is a possibility that this is wrong. This is how probabilities work."

A way FiveThirtyEight tried to do that was to change the way they describe their forecasts. Instead of saying that a candidate had an 80 percent chance of winning, they might say that he or she had a four-in-five chance of winning. Someone with a 20 percent chance of winning would have a one-in-five chance. It's gambling odds vernacular, which they hope will make more sense to readers.

The simple change is something Kristen Soltis Anderson, a Republican researcher and strategist and co-host of the podcast *The Pollsters*, thinks will be helpful to laypeople going forward. "It gave people a better sense of what these probabilities actually mean," she said, and helps shape understanding that an 85 percent chance is not a 100 percent chance. Even if we should have all intuitively known that, "just that simple language tweak did make it more sensible," she said.

Thinking About the Numbers in a New Way

FiveThirtyEight and RealClearPolitics are prominent sites that aggregate numerous polls to calculate their average. It involves much more than simple math. FiveThirtyEight's models consider numerous details like historical accuracy, a bounce after a party convention, and if a candidate is an incumbent, and they give more weight to polls they think are trustworthy.

Sites like those can be helpful if you want a quick idea of the horse race, said Karlyn Bowman, an expert on public polling and a data analyst at the American Enterprise Institute, a right-leaning think tank. ("Horse race" has been used to describe political contests since the nineteenth century.)

But Bowman and other polling experts caution that the focus on the horse race both limits and misunderstands the primary purpose of polls. "Most of what you're doing in a poll is helping to figure out strategy and to assess where movement may or may not come from. But there's so much attention on that horse race, and so little understanding of the uncertainty that surrounds it," said Anderson.

VOTING TIP: A quick warning, and maybe the most important advice about voting and polling: never let your candidate's supposed probability of winning or losing keep you from going to the polls.

"These probabilistic forecasts can give potential voters the impression that one candidate will win more decisively and may even lower the likelihood that they vote," wrote Solomon Messing, then a Pew fellow who worked with researchers from the University of Pennsylvania and Dartmouth on a 2018 study on election forecasts.

Numbers that show a candidate's probability of winning can leave voters believing the race is less competitive than they'd understand it to be if they looked at the actual percentage of the vote a candidate is projected to get. You want to temper your understanding of an 80 percent chance a candidate will win with the notion that they're projected to get 51 percent of the vote.

And just as with news consumption, it's also important to go to more than one place for information. "Look at a lot of polls. That's always useful in a campaign cycle . . . particularly when an issue is new," Bowman said. In any description of a poll, whether in a news report or a candidate's ad, there needs to be more than just one number—a number only matters if there's another number to give it context. If 50 percent of those who answered poll questions support candidate A, you need to know how many support candidate B. It could be a close race, or 25 percent could support candidate B and 25 percent could be undecided.

If you're aiming for true diligence, also compare the numbers to those found in previous polls on the same topic, and then remember to look back and compare when a new poll is released. If a candidate's favorability rating is reported to have dropped from one week to the next—say, from 50 percent to 40 percent—it can look like a big fall. Maybe it is. But was it actually 40 percent two weeks ago, and for the six weeks before that? Will it jump back up to 50 percent the following week because the 40 percent was just an anomaly? As the election nears, things are fluid. Again, what makes polls about elections tricky is that only one day—Election Day—matters.

Beware the Margin of Error

"Margin of error" is a technical-sounding phrase reporters and poll readers make sure to emphasize. It typically "describes the amount of variability we can expect around an individual candidate's level of support," Pew researcher Andrew Mercer explained in a report. In other words, it says by how much a poll might be off. He gets funny and a little philosophical when describing how we're supposed to understand and explain the margin of error's impact: "as is so often true in life, it's complicated."

What does it mean if a poll has 47 percent for candidate A and 45 percent for candidate B (leaving room for undecideds or third-party candidates) and a 3 percent margin of error? "It could be a dead heat; it could be that one candidate has a pretty substantial lead," Bowman said. The 3 percent means that either of the numbers could move up or down three. That could bring candidate B's 45 percent up to as high as 48 or down to as low as 42. And if candidate A's 47 is actually 50, that can mean an 8-point lead for candidate A. That's why when a race looks to be falling within a poll's margin of error, Bowman again suggests checking out other polls to see how they compare.

Don't Judge a Poll (Only) by Its Numbers

Now that we're clear on how to think about the numbers, we're all good with polls, right? Of course not! You also need to think about who conducted the poll, and how they did it. It's similar to evaluating a media outfit's trustworthiness: Do you recognize the name of the company that published the poll, and did they show their work by explaining how many people were polled, who they were, and how the people were contacted? Did they explain exactly what they asked? Was that work enough to convince you of its reliability?

Polls conducted by Pew and Gallup and those done by reputable news organizations are among those considered trustworthy. Major media institutions use what is sometimes called "gold standard" methodology to ensure that they have the right mix of people—and they have the money to invest in the polls, the *Washington Post*'s Bump said. (Conducting proper polls is a pricey enterprise.)

If you don't recognize the organization, look them up to see if they have a particular focus, angle, or specialty. You might not recognize the name Kaiser, for example, but the Kaiser Family Foundation website tells you that it's a nonprofit whose modern mission was established in the early 1990s. The foundation focuses on health care. Whether an organization releasing a poll is left-leaning or right-leaning is another thing to think about—the political orientation of an organization doesn't necessarily determine a poll's trustworthiness, but political bias does suggest that there is an effort being made to convince you of something.

Methods of polling can be via live telephone calls to landlines or cell phones, or automated calls—what the pollsters call IVR, or interactive voice response—where a recorded voice is used. Increasingly popular are online polls, though all online polls are certainly not created equal. Online polls should only be trusted if a reputable polling organization is reaching out to a carefully selected group of people. That's very different from a poll that someone puts on a website or Twitter, where anyone can participate. "If you know that you can opt-in to the poll, then that poll is not reliable," Bump said.

How the ideal poll is conducted is an increasingly complicated issue, one that pollsters themselves are struggling with. Polls used to be done primarily via phone call. But even after shifting from landlines to a mix of landlines and cell phones, response rates—people answering the phone and agreeing to participate—have dropped significantly. A low response rate doesn't mean a poll is inaccurate, but it does mean that the margin of error will be higher than if more people were reached.

No particular way of polling is necessarily superior, but evaluating how a poll was done in conjunction with who they were trying to reach is important. Check, for instance, the percentage of landline calls versus cell phone calls and consider whom the poll is surveying. If a poll was automated-voice, landline-call-only seeking to gauge opinion on a Democratic primary, "I'm not going to look at that at all," *The Pollsters'* Anderson said, because a poll like that will reach an older and whiter audience, and that is not the profile of a voter likely to participate in a Democratic primary.

Look at What the People Care About

If you want to ignore the horse race and instead focus on understanding what voters care about, look to issue polls. Bowman prefers those because they take the specific candidates out of the equation and show what is really on the minds of the American people.

Gallup has regularly asked Americans the same open-ended question since 1935: What do you think is the most important problem facing this country today? Polls like that "are really much more concerned about ordinary life, about what makes America tick," Bowman said.

A glance through the history of the Gallup Poll, via a *New York Times* interactive story, shows how much the answers reflect the news of the era. In 1935, when America was living through the Great Depression, a top concern cited was unemployment. ("Liquor control" and "soldiers' bonus" were also among the answers.) In March 1965, the month of President Johnson's famous civil rights speech, civil rights and race relations were indeed the primary concerns for Americans. The economy was the dominant issue in

Barack Obama's first months in office in 2009; the country had just experienced the collapse of major investment institutions and faced the possible collapse of the auto industry. Compare that to October 2019, when unemployment was low but the Trump administration was clashing with Congress over the impeachment investigation: "government/poor leadership" was the given answer of 34 percent of Americans, more than any other issue.

Polls are a crude and blunt instrument, but their ability to tell us more about ourselves is what makes them important in the overall political landscape, Bowman said. "They help us understand what a very complex and heterogeneous public is thinking or not thinking about a lot of different issues, and I think that's really important in our democracy."

CHAPTER TEN

EXPLAINING THE ELECTORAL COLLEGE

Protesters demonstrate outside the Pennsylvania State Capitol, Harrisburg, Pennsylvania, December 19, 2018 (Mark Makela/ Getty Images)

Only one Texan from U.S. House of Representatives District 36 could officially vote for Donald Trump for president: Republican elector Art Sisneros, a welding supply salesman from Liberty County. His power stemmed from the U.S. Constitution, which says that the Electoral College, not the people, officially selects the president. All the other Texans in District 36 who voted for Trump were actually voting for Sisneros to vote for Trump. The same idea is true for every American who casts a presidential ballot. We aren't voting for the candidate; we're telling the Electoral College what we want them to do.

Sisneros's fellow citizens in Liberty County (the very red county where I grew up) made crystal clear how they expected him to vote. Trump won Texas with 52 percent of the vote statewide and received 78 percent of the vote in Liberty County.

The electors meet in person to vote, usually in their respective state capitals. The designated day is the first Monday after the second Wednesday of December. (Simple!) But Sisneros had a problem. He didn't want to vote for Trump.

More than thirty electors gathered in the main legislative chamber of the mammoth pink limestone Texas Capitol in Austin on December 19, 2016. The observation gallery was full, the lights twinkled on a two-story Christmas tree, and both sunlight and the shouts of protesters streamed in from outside. After a local girls' choir sang the national anthem and a reverend said a prayer, the Texas secretary of state called roll, one elector at a time. The electors responded with "here" or "present," but when Sisneros's name was called, there was silence.

After a significant amount of soul searching, Sisneros had decided several weeks earlier that he couldn't in good conscience vote for Trump. He resigned his position as elector.

So what happened?! Electors are so important that the Constitution prescribes that they choose the president, so was there a full-out crisis in the state of Texas? Were there days of emotional and intellectual debate over whether Sisneros should be forced to show up and vote his conscience or be banned forever from the Capitol grounds for refusing to heed the will of nearly everyone in his home county?

Nope, none of that happened. The electors voted, with little fanfare, to replace Sisneros with a woman from another small town in the same congressional district who was happy to vote for Trump. All the electors voted by secret ballot.

As it turned out, two other electors also refused to toe the party line. One voted for longtime Texas representative Ron Paul and one for John Kasich, who was then the governor of Ohio. But the other thirty-six electors chose Trump, and the Texas secretary of state announced that Trump had enough electoral votes to be named president. The gallery cheered. A reporter for the *Texas Tribune* on the scene said that Trump had passed the number of Electoral College votes he needed about thirty minutes earlier, in another statehouse.

That whole scenario is representative of the complexities and paradoxes that surround the Electoral College. The process is largely a formality, and the individual electors—mostly local party officers or friends of politicos—have no role besides checking or writing down a name that's already been designated for them. What those who voted for someone other than Trump did is not especially common, but it's not unheard-of. There's even a term for them—"faithless electors." But protest votes like those have never had an impact on an election.

THE ELECTORAL COLLEGE AS ENTERTAINMENT

The Electoral College is complicated when it functions properly, but what happens if things *really* go awry? Two modern examples are from fictional television.

The one handled with calm and certitude befitting the demeanor of a president was on *The West Wing,* when President-elect Matthew Santos, played by the dashing Jimmy Smits, must name a vice president after his running mate dies on Election Day. The (television) consensus was that it was legal for him to simply ask the Electoral College to vote for his new nominee. Santos instead opts to ask the Senate to confirm his preferred VP; he wants both to earn political points and to avoid any questions of legitimacy. (The series ended after that, so we'll have to assume the Senate approved.)

In *Veep*, Selina Meyer, the candidate always *this close* to a legitimate presidency, finds herself tied with her opponent for Electoral College votes. When that happens, the vote for president goes to the House of Representatives—each state getting one vote—and the Senate chooses the vice president. If that sounds like it was written for TV, it isn't. That's really the way it would work. It didn't work out for Meyer, who couldn't get the votes in the House.

Electors as Mirrors, not Trustees

"Since sometime fairly early in the nineteenth century, electors have been expected to be mirrors, to be delegates rather than trustees," said Sanford Levinson, professor of government at the University of Texas at Austin law school and an expert on the Constitution. "Mirrors" means the electors are expected to vote for the state's popular-vote winner.

The president is chosen by the people in a popular vote, but instead of a nationwide tally, it's fifty-one popular-vote contests, one for each state and Washington, DC. "College" in this case means an organized group of people with a common task.

The larger the population of the state, the more electors it gets. The number is calculated by adding the number of each state's U.S. senators, which is two, to its number of seats in the U.S. House of Representatives. Each state therefore has a minimum of three electors. California currently has the most, with 55. There are 538 total, and a candidate needs 270 to win.

States have their own methods for choosing electors, but they are usually selected by members of the state's political party at that party's state convention or by vote otherwise. Usually they're chosen as a sort of reward for loyalty and service to the party, or maybe because they have an affiliation to the presidential candidate. (Bill Clinton was tapped to be a New York delegate for Hillary.) Both Republicans and Democrats have their own electors, though the electors for the candidates who didn't win in that particular state aren't needed.

In forty-eight states, the winner of the state popular vote gets all the electoral votes for that state. So whichever candidate gets the greatest number of votes in California gets all fifty-five electoral votes, even if it's a close race. In Maine and Nebraska, the two states that aren't winner-take-all, the popular-vote winner gets two electoral votes, and the rest go to the candidate who won the popular vote in each of the state's congressional districts. Unlike California or Texas, with their dozens of congressional districts, Maine has only two districts. Nebraska has three. While they're outliers in the calculations, neither has played a deciding role in a presidential race.

As of the 2016 election, twenty-nine states and Washington, DC, bound their electors to vote for the popular-vote winner in the state, some by formal state laws. Colorado is one of the states that binds electors by law. Clinton won the popular vote in Colorado, but a twenty-four-year-old Colorado Democratic elector voted for Kasich instead of Clinton. The idea was that if enough electors in Colorado and elsewhere voted for Kasich, he could overtake Trump. Few adopted the idea, and Colorado refused to count the vote. The elector sued.

In August 2019, the Tenth Circuit Court of Appeals ruled that removing the elector and refusing to count his vote was unconstitutional, thus calling into question the constitutionality of state laws that bind electors. The U.S. Supreme Court agreed to hear the case and a decision is expected before the 2020 presidential election. It's unclear if it will have much practical impact; most electors are chosen for their loyalty to the party and aren't likely to step out of line in large numbers. And, again, "faithless electors" haven't been a consequential issue in states where electors aren't bound by law.

THE ELECTION OF 1800

Originally, there was not a separate ballot for president and vice president, so the person who won the greatest number of Electoral College votes was president, and the second-place winner was vice president. In the 1800 election, presidential candidate Thomas Jefferson and running mate Aaron Burr ran against incumbent John Adams. When the Electoral College voted, Jefferson and Burr tied. Battling party affiliations meant that some preferred Burr, and it took a week and thirty-six rounds of votes in the House of Representatives to sort things out. Jefferson became president, with Burr as vice president after Alexander Hamilton supported Jefferson. (Burr was still VP when he shot and killed Hamilton in a duel.) The 1800 election led to the Constitution's Twelfth Amendment, which requires separate votes for president and vice president.

Why Do We Have the Electoral College?

The Electoral College was created at the 1787 Constitutional Convention in Philadelphia. Experts disagree on many aspects of why the drafters chose this method for presidential elections, but one reason most get behind is a practical one. While the idea of a popular vote was suggested, the drafters were concerned that the country wouldn't have an opportunity to learn enough about the candidates. In the 1780s, there was no system of mass communication. Further, counting and reporting the popular vote would be something that would have taken months and involved a lot of uncertainty and logistical challenges.

The Electoral College system is unique among democracies. Part of the reason it was created was to put distance between the executive and legislative branches by making sure that the president wasn't selected by legislators, the way the prime minister is chosen in Britain's parliamentary system. No sitting U.S. senator or representative can be an elector. The framers wanted the presidency to be a separate branch of government, and for lawmakers to select him or her only in the case of a tie in the Electoral College.

The certainty among experts on why the Electoral College was created doesn't extend too much beyond this. Some believe it was a way to save the country from a bad selection. *The Federalist Papers* was a group of essays sent to newspapers to encourage states to ratify the Constitution, and in Federalist Paper No. 68, Alexander Hamilton discusses the Electoral College, saying that the electors will perform one job, and that's to assemble "and vote for some fit person as President." The Electoral College, Hamilton wrote,

> *affords a moral certainty, that the office of President will never fall to the lot of any man who is not in an eminent degree endowed with the requisite qualifications. Talents for low intrigue, and the*

little arts of popularity, may alone suffice to elevate a man to the first honors in a single State; but it will require other talents, and a different kind of merit, to establish him in the esteem and confidence of the whole Union.

This seems like clear-cut evidence of where the founding fathers stood. But Hamilton was in salesman mode, urging adoption of the Constitution, and some prominent experts think that he could have been finding a compelling reason to justify the Electoral College that went beyond the framers' reasoning at the time they created it.

There is even disagreement over whether the Electoral College was designed to pacify framers from slaveholding states, many of whom owned slaves themselves. In a popular vote system, the larger number of eligible voters in Northern states meant Southern candidates were likely to be shut out. But the South's population was much greater than its number of eligible voters—it was also home to many enslaved people. In an Electoral College system based on population rather than the number of eligible voters, things would be more even between the North and South. In the end, slaves counted as three-fifths of a person.

But it's debatable whether this benefit for slaveholding states was a driving factor to create the Electoral College. "The Electoral College imported the compromises on slavery that had already been reached dealing with congressional representation," historian Alexander Keyssar said. "In that sense it did reward slaveholders, but I don't think it was the point of this particular weird institution."

Why Do We Still Have the Electoral College?

Whatever the reasons the framers had for creating the Electoral College, the practical ones have little relevance in today's world. We have immediate access to an endless amount of information about the candidates, so we don't have to worry about citizen access to information. And though some would still argue that the change to a popular-vote system could be a mess—a state requiring a recount could prevent certification of the result, for example—we currently trust state vote counts before the electors' vote and have rules and deadlines by which they must be complete.

States' rights and population concerns do still dominate arguments about why we should keep the Electoral College. If the presidential race were decided by popular vote, the argument goes, candidates would spend all their time in New York and Los Angeles and Houston and Chicago recruiting the huge numbers of voters in those top-population cities, and, once the primaries were over, never set foot in the cornfields of Iowa. And that might be true. But as it stands, once the primary elections are over, candidates mostly just stealthily stop by the biggest cities to raise money and then spend all their time in swing states, none of which are home to the millions of citizens in New York, Los Angeles, Houston, or Chicago—though with Texas approaching swing-state status, Houston may soon find itself quite popular for campaigns.

Swing states are those states whose popular-vote outcomes aren't so predictable. Their popular vote might go to a Democratic candidate in one election and a Republican the next, which means that their Electoral College votes are often up for grabs.

In the 2016 presidential election, only a few states were won by less than 2 percent. Clinton won New Hampshire and Minnesota. Trump won Florida, Michigan, Pennsylvania, and Wisconsin, all states that had gone for Obama in 2012. Michigan, Pennsylvania, and Wisconsin have 46 electoral votes, and if Clinton had won them, she would have surpassed 270 electoral votes and won the presidency.

But she didn't. More than 135 million people cast a vote for president in the 2016 election, but about 80,000 voters in Michigan, Pennsylvania, and Wisconsin decided the election.

THE ELECTION OF 1824

Four serious contenders for president led to an Electoral College debacle in 1824. Andrew Jackson won the greatest number of Electoral College votes, but not enough to win the presidency. The fourth-place Electoral College finisher, Henry Clay, "eagerly assumed the role of kingmaker," Norman Ornstein wrote in *After the People Vote: A Guide to the Electoral College.* Clay met with fellow candidate John Quincy Adams, and there may or may not have been an agreement that Clay would be secretary of state if he delivered the votes of Ohio and Kentucky, which Clay had won, to Adams. Clay denied that there was any quid pro quo, but Adams became president and Clay indeed became secretary of state. Four years later, Jackson, in a populist campaign that decried backroom politics, won the presidency.

Proposals to End the Electoral College

Trump won states that had a total of 306 electoral votes, pushing him well above what he needed. Clinton won the popular vote by a margin of almost 2.9 million votes. It was the fifth time in history that the popular-vote winner lost the presidency. A week after the election, U.S. senator Barbara Boxer of California proposed a constitutional amendment to abolish the Electoral College. It was far from the first such plan. Over time, there have been more than seven hundred proposed bills or amendments to abolish or modify the Electoral College process, according to FairVote.

Any debate among scholars and experts over whether or not the Electoral College still makes sense begins with the same disclaimer: the candidate who wins the Electoral College vote almost always also wins the popular vote. And, other than in 2016, the opposite has happened only once in recent history, when Al Gore beat George W. Bush in the popular vote in 2000. After losing the Supreme Court battle over the Florida recount, and thus that state's Electoral College votes, Vice President Al Gore had to oversee the official Electoral College count in his role as president of the Senate. There were objections to the counts on Gore's behalf, but not enough. George W. Bush took the oath of office three weeks later.

The 1900s were conflict-free with regard to the electoral vote versus the popular vote. Not so in the 1800s. In 1824, Andrew Jackson won the popular vote, but John Quincy Adams became president; in 1876, Samuel Tilden won the popular vote but Rutherford B. Hayes the presidency; and in 1888, incumbent Grover Cleveland was the popular-vote choice, while Benjamin Harrison won the Electoral College. (Of those elections, only Tilden never served as president.)

One reason that the people (you, me, and every other citizen) haven't taken to the streets over this issue is that it so rarely matters. When it does, we're told it's an anomaly. The Electoral College has remained intact since the early days of the nation, despite the fact that it's long been generally unpopular with a majority of Americans. (After the 2016 election, the Electoral College lost even more favor with Democrats, but gained Republican support.) One of the main reasons is that it's nearly impossible to change. The only real way to alter or end the Electoral College is to amend the Constitution. That's extremely difficult to do under the best circumstances, and impossible in a divided political climate.

To amend the Constitution, both the Senate and the House must approve the change by a two-thirds vote, and three-fourths of states must agree as well. As of early 2020, the Constitution has only twenty-seven amendments, and the last one was added in 1992 and involved congressional pay. Before that, in 1971, an amendment lowered the voting age to eighteen.

In semi-recent history, both the House and the Senate have passed bills to end or dramatically change the Electoral College, but they've never wanted to do it at the same time. In 1950, a Republican senator and a Democratic senator co-sponsored a proposed amendment that would have kept the Electoral College but changed it so that electoral votes would be cast in proportion to the state's popular vote. If the Republican candidate won 60 percent of the vote in a state and the Democrat 40 percent, the electoral votes for that state would have been split in that manner. It passed the Senate by a margin of 64 to 27, but the House showed no such enthusiasm.

The House was interested in a change in 1969, after independent presidential candidate and segregationist George Wallace secured forty-six Electoral College votes the year before. The House passed a proposed amendment that would have seen the president elected by popular count provided he or she won at least 40 percent of the vote. If no candidate reached that threshold, there would be a runoff. Despite its popularity in the House, where it passed by a margin of 339 to 70, the Senate didn't take the bill to a vote.

THE ELECTION OF 1876

The election of 1876 was ugly. Republican candidate Rutherford B. Hayes's campaign accused Samuel Tilden of being a drunkard, a swindler, and, in a description one doesn't see often these days, syphilitic. Tilden's side said that Hayes had stolen from dead soldiers and shot his own mother. Tilden received more popular votes, and was only one vote shy of the number of Electoral College votes he needed, with several states still counting.

The story goes that the national Republican Party leaders put pressure on the states still counting ballots, telegramming them to say that if their state went for Hayes, he would win. The national Republican Party chairman then claimed that Hayes had won, and all hell broke loose.

President Grant sent troops to keep the peace as the votes were actually counted, and accusations of cheating abounded. The dispute dragged on until March 1877. Because there was no plan for what to do about conflicting state returns, a bipartisan commission was created, but everyone voted along party lines and eventually Hayes was named president. Tilden believed until his dying day he should have been president.

One of the compromises to end this dispute was that the North pull all its troops out of the South. It signaled the end of Reconstruction and many of the political gains of African Americans in the South.

Small States, Swing States, and the Electoral College

Even if one of those proposals could have passed both the Senate and the House, any amendment would need approval by three-fourths of the states. But both smaller states and swing states enjoy an outsized power, and that power earns them a lot of attention from lawmakers and candidates.

"In 2000, George Bush got nine votes by carrying the two Dakotas and Wyoming. Al Gore got five votes by carrying New Mexico," constitutional law professor Levinson points out. New Mexico had roughly the same population as the other three states combined. It's a vivid, if selective, example of the difference it makes in giving North and South Dakota and Wyoming an Electoral College boost their population numbers don't warrant, Levinson continues. (He is among those who believe that the Electoral College has outlived its usefulness.)

Battleground states, like Florida and Ohio, have much larger populations and Electoral College representations, and thus also see the extra benefits and attention of being high-stakes, multi-visit states for candidates. As some states inch closer to battleground status, it's in the majority's interest to keep the Electoral College. The Texas statehouse is 55 percent Republican, and the state voted 52 percent for Trump in 2016. The current majority party would have no reason to lessen its Electoral College power by allowing the 3.9 million Texans' votes that went for Clinton to actually contribute to the outcome.

Trump won convincingly in twenty-six states, and thirty-eight states would need to approve a constitutional amendment. If the popular vote chose the president, Clinton would have won and a majority of citizens in those twenty-six states would not have seen their choice in the White House. To pass an amendment, a majority of those states would have to vote to eliminate their influence.

Keyssar, who is working on a book about the Electoral College, explains that despite the outsized influence small states have, it isn't necessarily those states that stand in the way of change. Some small state citizens and lawmakers, his research indicates, might vote to end or alter the Electoral College even to their own supposed detriment.

It's states like Texas and Florida, with a Republican majority losing ground to growing minority populations who usually vote Democrat, that are more likely to draw a hard line. Bottom line: between the political party splits and the swing-state power, the Electoral College is likely to stick around for a while.

In multiple interviews with scholars and political operatives, there was one question no one could answer. What happens if a split between the majority vote winner and the Electoral College winner becomes the norm? What if it's two elections in a row, or three? Will the majority voting public continue to accept it? The Constitution didn't plan for that, and politicians today have no plan for it, either.

I've Got a Pact for That

Citizens who would like to move to a popular-vote system aren't completely stuck, though it does still come down to how their state feels. In recent years, multiple states have signed into law the National Popular Vote Interstate Compact, which would require those states' electoral votes to go to the winner of the nationwide popular vote.

But there's a catch. The compact doesn't go into effect until enough states have passed it so that their electoral votes add up to 270. States don't want to risk giving away their electoral votes if they aren't aligned with the popular-vote winner unless others are doing the same. Maryland became the first state to pass the compact in 2007. In total, 15 states and Washington, DC, have joined. When Oregon joined, in 2019, the states in the compact had a combined 196 electoral votes.

John Fortier, editor and co-author of *After the People Vote*, cautions that things might not be so simple, even if enough states do join to get to 270 electoral votes. It's an untested system, so it's not clear what would happen if a state that had joined the compact backed out after an election. The 2019 federal appeals court decision indicating states couldn't bind their electors muddied the water a little, though Fortier predicted states that have signed it would just choose electors who pledge to honor the compact.

Unknowns aside, the compact is a simpler approach than trying to amend the Constitution, even if it still faces an uphill climb. To get the needed 270, some historically Republican-leaning states would have to sign on. Because the Electoral College is currently considered to benefit that party, it's unlikely they'd do so.

The Whole Point of It All? Show Up. It Matters.

So, how should you think about the Electoral College as you go to vote? Well, if you live in a battleground state, the candidates visit you a lot for a reason. Your vote is especially sought-after in presidential races and could help determine a close election. Get to the polls, and recruit people to join you.

And if you live in a state that's already pretty red or pretty blue? First, if you've learned anything from this book, I hope it's that the most important elections aren't always at the top of the ballot. The actions of your school board members affect every child in your community. Your mayor and city council members make decisions that touch your life every day. Your state representatives govern your health care and your voting rights, and may draw your U.S. congressional district lines. And the Electoral College has nothing to do with how you choose your U.S. representatives and senators.

Even if you're likely to be in the minority in your state's presidential race, it's still imperative that you vote. Candidates should know where state citizens stand. They are there to serve you, even if you don't always agree with them, or they with you. Voters for the losing party still have influence, especially if their numbers are mighty.

Plus, we are far from 100 percent turnout, which means there's room for improvement and increased voting power within every age group and racial group, and at every income level. That's especially true for groups with lower turnout rates, including young people, Hispanics, and Asian Americans. In some places and races, even moderate increases in turnout can decide the election.

Your vote is your voice. Use it. And then use your actual voice, your texting fingers, your Instagram images, and any other social media you've got to encourage the people you know and love to vote. You—not a politician, not a celebrity, not an athlete, but you—are the greatest inspiration to your family, your friends, your classmates, your coworkers, and your online community.

From me to you, sincerely: Thank you for voting.

THANK YOU FOR VOTING: A CHECKLIST

Below is a list of everything you need to do to be able to vote, some steps to take to be a confident and informed voter, and a few extra actions that will strengthen your own impact by making sure your friends and family vote. You might even consider volunteering for a candidate, ballot initiative campaign, or issue group.

TODAY

- ❒ Register to vote.
  - VOTE411.org, Vote.org, or TurboVote.org will help you register and follow the proper procedures for your state.
- ❒ Choose five friends to join you to vote. Set up a text chain and promise to hold each other accountable.
- ❒ Learn what choosing a particular party means in your state. Some states have closed primaries—only those registered to vote for that party will be able to vote in that party's primary elections.
- ❒ If you're already registered, confirm that your registration is up-to-date; if you've moved since you registered or updated, provide your correct address.
- ❒ Check what ID (if any) your state requires you to bring with you in order to vote.
  - USA.gov can connect you to your state's info, and VOTE411.org is also a good resource.
- ❒ If you don't already have proper ID, make a plan to get one.
  - VoteRiders.org is an organization that helps people obtain IDs.

- ❒ Sign up for election-related text alerts at TurboVote.org.
  - Your state, county, or city may offer a similar service. You'll get texts for registration deadlines, reminders of upcoming elections, and Election Day reminders to vote, all specific to you and your community.
- ❒ Mark your calendar with this year's elections, including local, state, and national races, and any primary elections to choose party candidates. (Don't rely solely on reminders!)
  - USA.gov will connect you to your state and local sites. VOTE411.org and Ballotpedia.org are also good resources for dates and races in upcoming elections.
- ❒ Check the dates for candidate debates for local, state, and national races. Consider inviting friends to watch the debates with you.
- ❒ Ask your employer if they provide time off to vote, and check state laws, too.
- ❒ Ask your employer if they have a plan to encourage employees to vote. If they do, offer to help get the word out. If they don't, consider leading the effort.
- ❒ If you'd like to volunteer to get out the vote for any candidate or ballot initiative campaign, contact that campaign, your county or state party association, or prominent party organizations for information.

40 DAYS BEFORE AN ELECTION

- ❒ Confirm that your registration is active and up-to-date. If you haven't registered, make sure to do so before your state deadline. (State cutoffs vary, but all states allow registration at least up to thirty days before federal elections.)
- ❒ Check VOTE411, Ballotpedia, and your state or local election websites, which should have updated information on what's on your ballot.

- ❒ Read news stories from a variety of sources about the candidates and issues to help you make informed votes.
- ❒ Review candidates' websites, and those of issues-based organizations and political organizations you trust.
- ❒ Text your five friends these steps.

30 DAYS BEFORE AN ELECTION

- ❒ Check VOTE411 or your state or local election information for the voting options in your state and county.
- ❒ Do you have the option to vote by mail?
    - If you're voting by mail, keep an eye out for your ballot, know when it must be mailed back, and make sure it's postmarked by the correct date.
- ❒ If in-person voting is standard, will you be in town on Election Day? If not, check the rules of voting absentee or early.
- ❒ Does your state have early voting? If so, check the early voting period and your polling location, which might be different from your usual polling place.
- ❒ Make a plan for when and how you'll vote, including what time and how you'll get there, and put it on your calendar. Set an alert to remind you.
- ❒ Text your five friends these steps.
- ❒ Make a plan to meet up with friends to vote if you share a polling place, or to meet after to celebrate having done your civic duty.

10 DAYS BEFORE AN ELECTION

- ❒ Do any additional research you need to help you make an informed choice.
- ❒ Host a "ballot party" to go over your ballot with friends, and work together to research any of the races you're uncertain about.

- ❒ Confirm your voting plan.
- ❒ Check if there are any #VoteTogether Election Day parties in your area.
- ❒ Ask a child or teenager you're close with to join you to vote.
- ❒ Text your five friends these steps.
- ❒ Share your voting plans and ask them theirs.

THE DAY BEFORE

- ❒ Confirm your polling location.
- ❒ Confirm that you have your proper ID, if necessary.
- ❒ Check the weather and plan accordingly.
- ❒ Text your five friends these steps.

ELECTION DAY (OR YOUR VOTING DAY)

- ❒ VOTE!
- ❒ Text your five friends to vote!
- ❒ Brag! Tell your friends you voted! Post photos on social media.

THANK YOU FOR VOTING: TELL YOUR FRIENDS

Choose five people you're going to make sure vote. Remind them of the steps along the way. Tag them below (the analog way—write it down!), then post the picture on social media, and tag them there, too. Include @thankyouforvoting on Instagram and #thankyouforvoting on Twitter.

Dear

@______________________________

@______________________________

@______________________________

@______________________________

@______________________________

Let's make our voices heard on Election Day. I pledge to you that I'm going to vote in the next election. Will you pledge to vote, too? Thank you for voting!

Sincerely,

@______________________________

ACKNOWLEDGMENTS

This book would not exist without Ann Patchett. What began as a research project for Ann became this writing adventure. For that opportunity, as well as her indispensable advice, endless encouragement, and honest opinions, I am forever grateful.

Gail Winston's editing made the book far better (and more concise!), and I appreciate so much the care she took with it. Dan Kirschen fielded many, many questions and hastily drafted emails from this first-time author. He handled them all with patience and humor. For those reasons, as well as his expertly delivered, dead-on editing suggestions, I offer a huge thank-you. Andrea Guinn designed a cover I truly love and Janet Byrne delivered a master copy edit. The photos wouldn't have been half as lovely without the help of Erica Singleton. Katherine Beitner and Erin Kibby attacked publicity and marketing with enthusiasm from the start.

Alicia Tan at HarperCollins and Andrianna Yeatts at ICM always had a ready answer no matter how basic or insane my questions. Jonathan Burnham was kind enough to listen to the very first sketches of what this book would become, and his belief in that early outline gave me the confidence to continue this project. Jane Beirn acted on my out-of-the-blue query, a kindness I'll never be able to repay.

Andy Young gave me my first internship in journalism. And now, years later, he graciously agreed to fact-check this book. It is infinitely better for it. May he always suggest Elizabeth Cady Stanton as a Halloween costume.

I especially want to thank every source who agreed to be interviewed. They answered my countless questions and taught me so much, and I always walked away inspired by their creativity and dedication to their work. They are champions of democracy.

So many friends happily and willingly suffered through pages for which "rough draft" was too kind a description. Kristina Peterson took the earliest pass at the first chapters. Christina Masso and I studied for news quizzes together in college, and when I sent her the full book and asked her to read it immediately, she was at the ready. Her brilliant editing saved me.

My switch to a career in journalism resulted in friendships I couldn't survive without. Thank you to these journalists and writers for their encouragement, editing, advice, and support: Rachel Dodes, Jamie Stelter, Alison Frankel, Kara Jesella, Danielle Friedman, and Candace Jackson. I admire them endlessly. Katie Rosman came through with the right advice and tough love at the exact right time(s). Elizabeth Holmes was kind enough to write a book at the same time—commiseration in the lowest lows was necessary, and I look forward to sharing the highest highs, too.

Lexi Mainland lived with this book nearly every day that I did. For every text I extend one thank-you—so a million or so. I spent many Central Park walks working through writing issues with Isabel Murphy; I'll repay the favor when it's time for her brilliant work to debut.

My journalism school "lunch bunch" is the support system of my dreams. Brad Davis always has the right words, whether of encouragement or about copy, and in addition to thoughtful suggestions, Esmé Deprez and Alex Lowther provided housing and transportation for reporting trips in California. Kristina Peterson and Isaac Baker did the same in Washington. Isaac gets a special thank-you for sending a very long list of source ideas in response to a very vague "I'm going to research voting" email. While I'm at it, the excitement my law school boys, Marc Ellenbogen and Alex Kaplan, expressed about this project was also very much appreciated.

Amy Penn welcomed me into her quiet, lovely home in my two-week sprint to finish the draft. Thank you for all services rendered, all these years. My first roommate, Adrienne Domas, was a light as well during those weeks, and Laura Lee Daigle was thankfully there too, even delivering a Sonic slush—a true Texas favor if there is one. Thank you to Erin Naman for what I think was the first pre-order. Emily Iverson—one huge thank-you for your many, many years of service. To Leslie Goodwin, the same!

Ashley Wu created and opened the most beautiful workspace; I love seeing you at your table when I turn the corner. The ladies and staff of Maison witnessed every step of this book and made it infinitely more pleasant (even when I wasn't). I'm so insanely grateful for the friendships Reed's preschool brought me—Lyn Devon, Heather Johnson, Chrissy O'Donovan—you're just the best.

I couldn't have done this without feeling like Reed was in good hands. A special thanks to Chenea Thorney for all the time she spent with him as I was writing the bulk of this book.

Myra Smith was the first professional writer I knew—I'm so grateful for my entire Smith family. The women of my family, Connie, Carri, and Ellis, are always endlessly supportive, no matter how many times I switch gears. I hope I've made Evan proud. My dad's enthusiasm for the book made me so happy. I owe my mom everything; I'm forever in awe of her unwavering love.

No one has brought more excitement about this book, nor marketed it better, than Reed. He is our kind, funny, curious everything. This book also would not exist without Bryan, who took on more than half the home-running without complaint, is brilliant and kind, and is always steady, no matter what. He is the best, full stop.

NOTES

Preface

ix "the only way they could do that is by not voting": Franklin D. Roosevelt—"The Great Communicator," The Master Speech Files, 1898, 1910–1945, File No. 1539, October 5, 1944, http://www.fdrlibrary.marist.edu/_resources/images/msf/msfb0170.

x "white primary": *Smith v. Allwright*, U.S. Supreme Court, 321 US 649 (1944), https://supreme.justia.com/cases/federal/us/321/649/.

xii Each generation votes: United States Election Project, http://www.electproject.org/home/voter-turnout/demographics.

xiv than Japanese citizens have in a lifetime: Donald P. Green and Alan S. Gerber, "Introduction: Why Voter Mobilization Matters," in *Get Out the Vote: How to Increase Voter Turnout*, 4th ed. (Washington, DC: Brookings Institution Press, 2019), chap. 1, Kindle.

xiv President Oops, Didn't Choose One: Philip Bump, "A Lot of Nonvoters Are Mad at the Election Results. If Only There Were Something They Could Have Done!" *Washington Post*, November 16, 2016, https://www.washingtonpost.com/news/the-fix/wp/2016/11/16/a-lot-of-non-voters-are-mad-at-the-election-results-if-only-there-was-something-they-could-have-done/.

xv "You pledged to vote": Donald Green, interview with author, April 24, 2019.

Chapter One: Democracy in Name Only

4 party-like atmosphere: Green and Gerber, "Evidence Versus Received Wisdom," in *Get Out the Vote*, chap. 2.

5 "New claims will arise": John Adams to James Sullivan, May 26, 1776, National Archives, Founders Online, https://founders.archives.gov/documents/Adams/06-04-02-0091.

5 From 1819 forward: Alexander Keyssar, *The Right to Vote* (New York: Basic Books, 2000), 44–45.

6 "Appeal of Forty Thousand Citizens, Threatened with Disfranchisement": Robert Purvis, "Appeal of Forty Thousand Citizens, Threatened with Disfranchisement," http://digitalhistory.hsp.org/pafrm/doc/appeal.

7 "The evolution of democracy": Keyssar, *The Right to Vote*, xxiii.

8 African Americans from Tennessee: Black Residents of Nashville to the Union Convention of Tennessee (1865), http://www.freedmen.umd.edu/tenncon.htm.

8 "put down their tools and ceased working": Keyssar, *The Right to Vote*, 73.

8 "Black political power": Blain Roberts and Ethan Kytle, "When the South Was the Most Progressive Region in America," *The Atlantic*, January 17, 2018, https://www.theatlantic.com/politics/archive/2018/01/when-the-south-was-the-most-progressive-region-in-america/550442/.

9 a federal suffrage amendment was a necessity: Keyssar, *The Right to Vote*, 74.

9 "opponents of a broad amendment": Keyssar, *The Right to Vote*, 81.

10 in the correct box: Darryl Paulson, "Florida's History of Suppressing Blacks' Votes," *Tampa Bay Times*, October 11, 2013, https://www.tampabay.com/news/perspective/floridas-history-of-suppressing-blacks-votes/2146546/.

10 turnout dropped from 62 percent to 11 percent: Michael J. Klarman, *From Jim Crow to Civil Rights: The Supreme Court and the Struggle for Racial Equality* (New York: Oxford University Press, 2004), 32.

10 down to around 5,300: James Bryce, *The American Commonwealth*, vol. 2, 3rd ed. (New York: Macmillan, 1897), 545, https://books.google.com/books/about/The_American_Commonwealth_The_party_syst.html?id=5K7ff2MCKBwC.

10 "purchase the presidency": Charles W. Calhoun, *Benjamin Harrison: The American Presidents Series: The 23rd President, 1889-1893* (New York: Henry Holt, 2005), 55.

11 certifying vote counts: Keyssar, *The Right to Vote*, 86.

11 By 1940, only 3 percent: "Voting Rights Act: Major Dates in History," ACLU, https://www.aclu.org/voting-rights-act-major-dates-history.

11 Lodge bill had proposed seventy-five years before: Keyssar, *The Right to Vote*, 210.

11 "shorthand . . . for describing African Americans": Becky Little, "Who Was Jim Crow?," *National Geographic*, August 6, 2015, https://www.nationalgeographic.com/news/2015/08/150806-voting-rights-act-anniversary-jim-crow-segregation-discrimination-racism-history/.

12 As early as 1887: "The Georgia Railroad Demands Relief of the Commission," *New York Times*, July 30, 1887, https://timesmachine.nytimes.com/timesmachine/1887/07/30/100925392.html?pageNumber=3.

12 black people in Tennessee were frustrated: "Breaking the Jim Crow Law," *New York Times*, July 30, 1915, https://timesmachine.nytimes.com/timesmachine/1915/07/30/104650796.html?pageNumber=8.

12 "separate but equal": *Plessy v. Ferguson*, U.S. Supreme Court, 163 U.S. 573 (1896), https://supreme.justia.com/cases/federal/us/163/537/#tab-opinion-1917401.

13 Mississippi's literacy test: *Williams v. Mississippi*, U.S. Supreme Court, 170 U.S. 213 (1898), https://supreme.justia.com/cases/federal/us/170/213/.

13 require voters to pay a poll tax: *Breedlove v. Suttles*, U.S. Supreme Court, 302 U.S. 277 (1937), https://supreme.justia.com/cases/federal/us/302/277/.

13 ended "separate but equal": *Brown v. Board of Education*, U.S. Supreme Court, 347 U.S. 483 (1954), https://supreme.justia.com/cases/federal/us/347/483/.

13 Student Nonviolent Coordinating Committee: Ari Berman, *Give Us the Ballot* (New York: Picador, 2015), 4–5.

14 a toothbrush, and two books: Berman, *Give Us the Ballot*, 5.

14 Lewis would weep: Rachael Bade, "John Lewis's Tears over Ancestor's Voter Card Stir Emotions in Democratic Caucus," *Washington Post*, June 12, 2019, https://www.washingtonpost.com/powerpost/john-lewiss-tears-over-ancestors-voter-card-stirs-emotions-in-democratic-caucus/2019/06/11/8f568332-8c60-11e9-adf3-f70f78c156e8_story.html.

14 "god-damnedest toughest": Edwin M. Yoder Jr., "Civil Rights: Much More to Do," *Washington Post*, May 19, 1982, https://www.washingtonpost.com/archive/politics/1982/05/19/civil-rights-much-more-to-do/b5a109f5-3b4f-4361-9aef-66e75c3baa45/.

14 "should be no argument": Lyndon Baines Johnson, "Special Message to Congress," LBJ Presidential Library, Speeches and Films, March 15, 1965, http://www.lbjlibrary.org/lyndon-baines-johnson/speeches-films/president-johnsons-special-message-to-the-congress-the-american-promise.

14 "brilliantly framed the cause": Berman, *Give Us the Ballot*, 27.

15 250,000 new black voters: "Voting Rights Act: Major Dates in History," ACLU, https://www.aclu.org/voting-rights-act-major-dates-history.

15 59.8 percent after: U.S. Commission on Civil Rights, "The Voting Rights Act: Ten Years After," January 1975, p. 43, https://lccn.loc.gov/75601086.

16 they could take power: Daniel McCool, Susan M. Olson, and Jennifer L. Robinson, *Native Vote* (Cambridge: Cambridge University Press, 2007), 3–4.

17 Arizona Supreme Court case: McCool, *Native Vote*, 15–16.

17 "we can't vote?": McCool, *Native Vote*, 10.

18 other taxes paid: McCool, *Native Vote*, 12.

18 the Clinton administration: Andrew Oxford, "It's Been 70 Years Since Court Ruled Native Americans Could Vote in New Mexico," *Santa Fe New Mexican*, August 2, 2018, https://www.santafenewmexican.com/news/local_news/it-s-been-years-since-court-ruled-native-americans-could/article_d0544a48-ef37–56ef-958f-eb81dcf01344.html.

18 western state elections in the 1950s: McCool, *Native Vote*, 20.

18 repealed the law the next year: McCool, *Native Vote*, 96–97.

18 terms favorable to them: McCool, *Native Vote*, 46.

19 Native Americans were persuasive: Daniel McCool, interview with author, June 25, 2019.

19 before they became citizens: Keyssar, *The Right to Vote*, 65.

20 "I know nothing": Lorraine Boissoneault, "How the 19th-Century Know Nothing Party Reshaped American Politics," Smithsonian.com, January 26, 2017, https://www.smithsonianmag.com/history/immigrants-conspiracies-and-secret-society-launched-american-nativism-180961915/.

20 write their names: Keyssar, *The Right to Vote*, 68.

20 "whites only" voting law in 1857: Oregon Secretary of State, "Crafting the Oregon Constitution," https://sos.oregon.gov/archives/exhibits/constitution/Pages/during-race.aspx.

20 bar natives of China from voting: Keyssar, *The Right to Vote*, 114.

21 Chinese immigrants given the opportunity: "Timeline of Chinese Immigration to the United States," The Bancroft Library, University of California Berkeley, https://bancroft.berkeley.edu/collections/chinese-immigration-to-the-united-states-1884-1944/timeline.html.

21 Japanese and other Asian immigrants: "Asian American History," Japanese American Citizens League, https://jacl.org/asian-american-history/.

21 half of eligible Asian Americans did: Jens Manuel Krogstad and Mark Hugo Lopez, "Black Voter Turnout Fell in 2016, Even as a Record Number of Americans Cast a Ballot," Pew Research Center, May 12, 2017, https://www.pewresearch.org/fact-tank/2017/05/12/black-voter-turnout-fell-in-2016-even-as-a-record-number-of-americans-cast-ballots/.

21 surge over the 2014 midterms: William H. Frey, "2018 Turnout Rose Dramatically for Groups Favoring Democrats, Census Confirms," Brookings, May 2, 2019, https://www.brookings.edu/research/2018-voter-turnout-rose-dramatically-for-groups-favoring-democrats-census-confirms/.

22 in order to assure the VRA extension: Keyssar, *The Right to Vote*, 225–27.

22 lower the voting age only for federal elections: *Oregon v. Mitchell*, U.S. Supreme Court, 400 U.S. 112 (1970), https://supreme.justia.com/cases/federal/us/400/112/.

Chapter Two: Long-Suffering for Women's Suffrage

24 170 meetings the year before to build the movement: Susan B. Anthony, diary entry, January 1, 1872, Susan B. Anthony Papers, Library of Congress, https://www.loc.gov/exhibitions/women-fight-for-the-vote/about-this-exhibition/seneca-falls-and-building-a-movement-1776–1890/a-movement-at-odds-with-itself/relentless-travel-and-a-new-departure/.

25 pass ratification there: Carrie Chapman Catt, Ratification Notebook, 1919–1920, Carrie Chapman Catt Papers, Library of Congress, https://www.loc.gov/item/prn-19–030/archival-materials-of-suffragist-carrie-chapman-catt-now-online/2019–03–18/.

25 "no voice, or Representation": Abigail Adams to John Adams, letter, March 31, 1776, Adams Family Papers, Massachusetts Historical Society, https://www.masshist.org/digitaladams/archive/doc?id=L17760331aa&hi=1&query

=%22declared%20an%20independency%22&tag=text&archive=letters&rec=5&start=0&numRecs=5.

26 "repeal our Masculine systems": John Adams to Abigail Adams, letter, April 14, 1776, Adams Family Papers, Massachusetts Historical Society, https://www.masshist.org/digitaladams/archive/doc?id=L17760414ja&hi=1&query=Intimation%20that%20another%20Tribe&tag=text&archive=letters&rec=2&start=0&numRecs=329.

26 Kentucky in 1838: "Kentucky and the 19th Amendment," National Park Service, https://www.nps.gov/articles/kentucky-and-the-19th-amendment.htm.

27 women's rights events: Sally Roesch Wagner, *The Women's Suffrage Movement* (New York: Penguin Books, 2019), 45.

27 women's rights convention was born: Wagner, *The Women's Suffrage Movement*, 57–58.

27 "the civil and political rights of women": Notice of Women's Rights Convention, *Seneca County Courier*, July 14, 1848, Library of Congress, available at https://www.loc.gov/resource/rbnawsa.n7548/?st=text.

27 The convention's Declaration of Sentiments: Liz Robbins and Sam Roberts, "Early Feminists Issued a Declaration of Independence. Where Is It Now?," *New York Times*, February 9, 2019, https://www.nytimes.com/interactive/2019/02/09/nyregion/declaration-of-sentiments-and-resolution-feminism.html; Valerie Jablow, "Tea and Sisterhood," Smithsonian.com, October 1998, https://www.smithsonianmag.com/history/tea-and-sisterhood-158244677/.

28 argued that suffrage: Ta-Nehisi Coates, "Frederick Douglass: 'A Women's Rights Man,'" *The Atlantic*, September 30, 2011.

28 eloquently about women's rights: Angela P. Dodson, *Remember the Ladies* (New York: Center Street/Hachette, 2017), 141–44.

28 "but it is comin'": Proceedings of the Woman's Rights Convention, September 6 and 7 (New York: Fowler and Wells, 1853), https://cdn.loc.gov/service/rbc/rbnawsa/n8289/n8289.pdf.

29 Their formal organization: Lori D. Ginzberg, *Elizabeth Cady Stanton: An American Life* (New York: Hill and Wang, 2009), 108–9.

29 "the Negro's hour": Dodson, *Remember the Ladies*, 191.

30 American Woman Suffrage Association: Allison Lange, "Suffragists Organize: American Woman Suffrage Association," National Women's History Museum, http://www.crusadeforthevote.org/awsa-organize.

31 "court their support for the ballot": Dodson, *Remember the Ladies*, 249.

31 upsetting for Southerners: Dodson, *Remember the Ladies*, 256.

31 telling his wife about the events of the day: Frederick Douglass obituary, *New York Times*, February 20, 1895, https://www.nytimes.com/2019/02/14/obituaries/frederick-douglass-dead-1895.html.

31 "abolitionists in disguise": Dodson, *Remember the Ladies*, 258.

32 "little band of nine ladies": "Minor Topics," *New York Times*, November 6, 1872, https://timesmachine.nytimes.com/timesmachine/1872/11/06/issue.html.

33 encourage women to move there: Mary Schons, "Woman Suffrage," *National Geographic*, January 21, 2011, https://www.nationalgeographic.org/news/woman-suffrage/.

33 supported their actions: "Utah and the 19th Amendment," National Park Service, https://www.nps.gov/articles/utah-women-s-history.htm.

34 independent of their husbands': Dawn Langan Teele, "How the West Was Won: Competition, Mobilization, and Women's Enfranchisement in the United States," *Journal of Politics* 80, no. 2 (April 2018): 442–61.

34 "We want our beer": Teele, "How the West Was Won: Competition, Mobilization, and Women's Enfranchisement in the United States," 447.

35 "we won't have it": Carrie Chapman Catt and Nettie Rogers Shuler, *Woman Suffrage and Politics* (New York: Charles Scribner's Sons, 1923), 89.

35 classist and anti-immigrant language: Keyssar, *The Right to Vote*, 164–65.

36 British suffragettes: Dodson, *Remember the Ladies*, 277.

36 Iowa and California: "Youngest Parader in New York City Suffrage Parade," Shall Not Be Denied: Women Fight for the Vote, Library of Congress, https://www.loc.gov/resource/ppmsca.58365/.

36 "moved by seeing marching groups": Harriot Stanton Blatch, "Why Suffragists Will Parade on Saturday," *New York Tribune*, May 3, 1912, https://chroniclingamerica.loc.gov/lccn/sn83030214/1912-05-03/ed-1/seq-1/. Dodson, *Remember the Ladies*, 279.

36 reintroduce the idea: Dodson, *Remember the Ladies*, 281–82;

36 "totally abhorrent": Elizabeth Cobbs, "Op-Ed: Woodrow Wilson's Woman Problem, a Case Study for the Trump Era," *Washington Post*, January 18, 2017, https://www.latimes.com/opinion/op-ed/la-oe-cobbs-wilson-womens-march-20170118-story.html.

37 "they made it their own": Janice Ruth, interview with author, May 30, 2019.

37 "heart of federal Washington": John Kelly, "Long Before Pink Hats, Female Protesters Marched in D.C. for Women's Rights," *Washington Post*, March 13, 2018, https://www.washingtonpost.com/local/long-before-pink-hats-female-protesters-marched-in-dc-for-womens-rights/2018/03/13/4b1335f2-26be-11e8-874b-d517e912f125_story.html.

37 "Enfranchising the Women": John Kelly, "Long Before Pink Hats, Female Protesters Marched in D.C. for Women's Rights," *Washington Post*, March 13, 2018, https://www.washingtonpost.com/local/long-before-pink-hats-female-protesters-marched-in-dc-for-womens-rights/2018/03/13/4b1335f2-26be-11e8-874b-d517e912f125_story.html.

38 "color prejudice so prevalent": Alice Stone Blackwell to Alice Paul, January 23, 1913, letter, Library of Congress, https://www.loc.gov/exhibitions

/women-fight-for-the-vote/about-this-exhibition/new-tactics-for-a-new-generation-1890–1915/new-tactics-and-renewed-confrontation/parade-planning-exposes-racial-divides/.

40 uphill climb: Dodson, *Remember the Ladies*, 290–91.

40 "can afford a little while to wait": Woodrow Wilson, "Address at the Women's Suffrage Convention, Atlantic City, N.J., September 8, 1916," President Wilson's State Papers and Addresses, 327, https://books.google.com/books?id=u2suAAAAIAAJ&pg=PA327&lpg=PA327&dq=wilson+%22I+have+come+to+congratulate+you+that+there+was+a+force%22&source=bl&ots=Ep78e_76_P&sig=ACfU3U2P85KPYHWHMpIti1XuR9ziPQsSUA&hl=en&sa=X&ved=2ahUKEwjatZDz2MTmAhUNTt8KHdHjB9cQ6AEwAXoECAsQAQ#v=onepage&q=%22I%20have%20come%20to%20congratulate%20you%20that%20there%20was%20a%20force%22&f=false.

40 "we want it to come in your administration": Ida Husted Harper, ed., National American Woman Suffrage Association's *The History of Woman Suffrage, Volume V: 1900–1920, After Seventy Years Came the Victory* (New York: J. J. Little & Ives Company, 1922), 488–99.

41 the vote instead of jail sentences: "Suffragists Wire Wilson," *New York Times*, July 19, 1917, https://timesmachine.nytimes.com/timesmachine/1917/07/19/96257621.html?pageNumber=2.

41 "'votes for women' sentiment": "Wilson, Shocked at Jailing Militants, May Advocate 'Votes for Women' as Part of War Emergency Program," *New York Times*, July 19, 1917, https://timesmachine.nytimes.com/timesmachine/1917/07/19/issue.html.

41 three-hour procession, Dodson, *Remember the Ladies*, 308.

41 Timothy Sullivan: Alice Sparberg Alexiou, "Tammany Hall, Women's Suffrage, and Big Tim Sullivan," Gotham Center for New York City History, November 10, 2017, https://www.gothamcenter.org/blog/tammany-hall-womens-suffrage-and-big-tim-sullivan.

42 "Jack Daniel's Suite": Jean Zimmerman, "Stirring, Engrossing 'Woman's Hour' Recounts the Battle for Suffrage," NPR.org, March 6, 2018, https://www.npr.org/2018/03/06/590072266/stirring-engrossing-womans-hour-recounts-the-battle-for-suffrage.

43 "some bunch of guys": Cokie Roberts, "100 Years Ago This Week, House Passes Bill Advancing 19th Amendment," interview by Steve Inskeep, NPR, May 22, 2019, https://www.npr.org/2019/05/22/725610789/100-years-ago-this-week-house-passes-bill-advancing-19th-amendment.

43 black women who had gone door-to-door: Jen Rice, "How Texas Prevented Black Women From Voting Decades After the 19th Amendment," Houston Public Media, June 28, 2019, https://www.houstonpublicmedia.org/articles/news/in-depth/2019/06/28/338050/100-years-ago-with-womens-suffrage-black-women-in-texas-didnt-get-the-right-to-vote/.

44 "sixty years of hard struggle": "Miss Susan B. Anthony Died This Morning," *New York Times*, March 13, 1906, https://timesmachine.nytimes.com/timesmachine/1906/03/13/101769455.html?pageNumber=1.

44 every presidential election since: "Gender Differences in Voter Turnout," Center for American Women and Politics, Rutgers University, https://www.cawp.rutgers.edu/sites/default/files/resources/genderdiff.pdf.

Chapter Three: Voting Problems and Voting Solutions

45 a 2017 Pew Study: "Public Supports Aim of Making It 'Easy' for All Citizens to Vote," Pew Research Center, June 28, 2017, https://www.people-press.org/2017/06/28/public-supports-aim-of-making-it-easy-for-all-citizens-to-vote/.

46 61 percent of eligible Americans voted: "Voting in America: A Look at the 2016 Presidential Election," Census.gov, Census Blogs, May 10, 2017, https://www.census.gov/newsroom/blogs/random-samplings/2017/05/voting_in_america.html.

46 shouting in the streets with delight: "Voter Turnout Rates Among All Voting Age and Major Racial and Ethnic Groups Were Higher Than in 2014," Census.gov, Census Blogs, April 23, 2019, https://www.census.gov/library/stories/2019/04/behind-2018-united-states-midterm-election-turnout.html.

47 "None of you would be here": Jeremy Bird, interview with author, March 16, 2018.

47 "varying election laws of the different States": James Bryce, *The American Commonwealth*, Volume II, 3rd ed. (New York: Macmillan, 1897), 142, https://books.google.com/books/about/The_American_Commonwealth_The_party_syst.html?id=5K7ff2MCKBwC.

48 register African American voters: Jonathan Mattise, Associated Press, and Elaina Sauber, *The Tennessean*, "Federal Judge Blocks Tennessee Voter Registration Law, Citing Harm to 'Constitutional Rights,'" *The Tennessean*, September 12, 2019, https://www.tennessean.com/story/news/2019/09/12/tennessee-voter-registration-law-blocked-judge-citing-harm/2300293001/.

49 prior felony conviction: Jean Chung, "Felony Disenfranchisement: A Primer," The Sentencing Project, June 27, 2019, https://www.sentencingproject.org/publications/felony-disenfranchisement-a-primer/.

49 voting in Florida in 2016: Tim Elfrink, "The Long, Racist History of Florida's Now-Repealed Ban on Felons Voting," *Washington Post*, November 7, 2018, https://www.washingtonpost.com/nation/2018/11/07/long-racist-history-floridas-now-repealed-ban-felons-voting/.

49 enthusiasm for second chances: Dale Ho, interview with author, July 23, 2018.

51 harder to vote: Carol Anderson, "Voter Suppression in the Twenty-First Century," interview by David Remnick, *The New Yorker: Politics and More*, podcast, December 3, 2018, https://www.newyorker.com/podcast/political-scene/voter-suppression-in-the-twenty-first-century.

51 with no available key: Mimi Marziani, interview with author, April 16, 2019.

53 Ray Price wrote: Berman, *Give Us the Ballot*, 85.

53 affecting black voters: Berman, *Give Us the Ballot*, 95–98.

53 Texas state agency: Berman, *Give Us the Ballot*, 106–13.

54 wasn't to be discriminatory: Berman, *Give Us the Ballot*, 145, 155–57.

54 she told lawmakers: Berman, *Give Us the Ballot*, 138–39.

54 "As I've said before": Howell Raines, "Voting Rights Act Signed by Reagan," *New York Times*, June 30, 1982, https://www.nytimes.com/1982/06/30/us/voting-rights-act-signed-by-reagan.html.

55 reported the *New York Time*'s Adam Liptak: Adam Liptak, "Supreme Court Invalidates Key Part of Voting Rights Act," *New York Times*, June 25, 2013, https://www.nytimes.com/2013/06/26/us/supreme-court-ruling.html.

55 "sue them over and over again": Daniel McCool, interview with author, June 25, 2019.

56 Hours after: Ed Pilkington, "Texas Rushes Ahead with Voter ID Law after Supreme Court Decision," *The Guardian*, June 25, 2013, https://www.theguardian.com/world/2013/jun/25/texas-voter-id-supreme-court-decision.

56 North Carolina also: Michelle Miller, Phil Hirschkorn, "Voter ID Bill Raises Controversy in North Carolina," CBSNews.com, August 13, 2013, https://www.cbsnews.com/news/voter-id-bill-raises-controversy-in-north-carolina/.

56 the final word: Camila Domonoske, "Supreme Court Declines Republican Bid to Revive North Carolina Voter ID Law," NPR, May 15, 2017, https://www.npr.org/sections/thetwo-way/2017/05/15/528457693/supreme-court-declines-republican-bid-to-revive-north-carolina-voter-id-law.

56 "sordid history": Martha Waggoner, "Sordid History Cited as Judge Blocks NC's Voter ID Law," *Washington Post*, December 31, 2019, https://www.washingtonpost.com/national/sordid-history-cited-as-judge-blocks-ncs-voter-id-law/2019/12/31/f3a41202-2c1d-11ea-bffe-020c88b3f120_story.html.

57 ending same-day registration options: "Election 2012: Voting Laws Round-up," Brennan Center for Justice, October 11, 2012, https://www.brennancenter.org/our-work/research-reports/election-2012-voting-laws-roundup.

57 American public favors: Kristen Bialik, "How Americans View Some of the Voting Policies Approved at the Ballot Box," Pew Research Center, November 15, 2018, https://www.pewresearch.org/fact-tank/2018/11/15/how-americans-view-some-of-the-voting-policies-approved-at-the-ballot-box/.

57 seventeen of those requiring a photo ID: "Voter Identification Laws by State," Ballotpedia, https://ballotpedia.org/Voter_identification_laws_by_state.

58 more than seven hundred thousand Texans: Issie Lapowsky, "A Dead-Simple Algorithm Reveals the True Toll of Voter ID Laws," *Wired*, January 1, 2018, https://www.wired.com/story/voter-id-law-algorithm/.

58 professor of political science at the University of Pennsylvania: Dan Hopkins, "What We Know About Voter ID Laws," FiveThirtyEight, August 21, 2018, https://fivethirtyeight.com/features/what-we-know-about-voter-id-laws/.

58 very little voter fraud to stop: German Lopez, "A New Study Finds Voter ID Laws Don't Reduce Voter Fraud—or Voter Turnout," Vox, February 21, 2019, https://www.vox.com/policy-and-politics/2019/2/21/18230009/voter-id-laws-fraud-turnout-study-research.

59 2006–2008 and 2014–2016: Kevin Morris and Myrna Pérez, "A Growing Threat to the Right to Vote," Brennan Center for Justice, July 20, 2018, https://www.brennancenter.org/our-work/research-reports/purges-growing-threat-right-vote.

59 "higher rates than other counties": Kevin Morris, "Voter Purge Rates Remain High, Analysis Finds," Brennan Center for Justice, August 1, 2019, https://www.brennancenter.org/our-work/analysis-opinion/voter-purge-rates-remain-high-analysis-finds.

59 cull their lists: *Husted v. A. Philip Randolph Institute*, U.S. Supreme Court, 584 U.S. __ (2018), https://supreme.justia.com/cases/federal/us/584/16-980/.

59 "notice from election officials": Adam Liptak, "Supreme Court Upholds Ohio's Purge of Voting Rolls," *New York Times*, June 11, 2018, https://www.nytimes.com/2018/06/11/us/politics/supreme-court-upholds-ohios-purge-of-voting-rolls.html.

60 "exact match": Miriam Valverde, "Georgia's 'Exact Match' Law and the Abrams-Kemp Governor's Election, Explained," PolitiFact, October 19, 2018, https://www.politifact.com/georgia/article/2018/oct/19/georgias-exact-match-law-and-its-impact-voters-gov/.

60 launched Fair Fight 2020: Greg Bluestein, "Updates: Stacey Abrams Launches National Expansion of Voting Rights Group," *Atlanta Journal-Constitution*, August 13, 2019, https://www.ajc.com/blog/politics/stacey-abrams-expanding-voting-rights-group-nationally/eeper6tpb1qNSUsZiOkHYJ/.

61 four-hour line: Celina Stewart and Jeanette Senecal, interview with author, March 27, 2019.

62 five actually cast a ballot: Jessica Huseman, "How the Case for Voter Fraud Was Tested—and Utterly Failed," ProPublica, June 19, 2018, https://www.propublica.org/article/kris-kobach-voter-fraud-kansas-trial.

63 the *Texas Tribune* reported: Emma Platoff, "Federal Judge Directs More Counties to Halt Voter Citizenship Review Efforts as Lawsuits Proceed," *Texas Tribune*, February 25, 2019, https://www.texastribune.org/2019/02/25/judge-tells-more-counties-not-purge-voters-now/.

63 "what's on my ballot?": Senecal, interview with author, March 27, 2019.

64 "helping people get to information": Senecal, interview with author, March 27, 2019.

64 experiment was successful: Steve Mistler, "Future of Maine's Ranked-Choice Voting Experiment at Stake in Tuesday's Election," Maine Public Radio, June 12, 2018, https://www.mainepublic.org/post/future-maines-ranked-choice-voting-experiment-stake-tuesdays-election.

66 behavior and voting laws: Paul Gronke, interview with author, March 6, 2019.

67 AVR voters in 2016: Rob Griffin, Paul Gronke, Tova Wang, and Liz Kennedy, "Who Votes with Automatic Voter Registration?," Center for American Progress, June 7, 2017, https://www.americanprogress.org/issues/democracy/reports/2017/06/07/433677/votes-automatic-voter-registration/.

68 in the time period observed: Kevin Morris and Peter Dunphy, "AVR Impact on State Voter Registration," Brennan Center for Justice, April 11, 2019, https://www.brennancenter.org/our-work/research-reports/avr-impact-state-voter-registration.

68 "AVR can do a lot in a state like that": Kevin Morris, interview with author, September 13, 2019.

68 Registration doesn't guarantee voting: Nathaniel Rakich, "What Happened When 2.2 Million People Were Automatically Registered to Vote," FiveThirtyEight, October 10, 2019, https://fivethirtyeight.com/features/what-happened-when-2-2-million-people-were-automatically-registered-to-vote/.

68 one million new voters registered: *Times* Editorial Board, "Despite Bungled Debut of 'Motor Voter' Law, It's Delivering New Voters as Promised," *Los Angeles Times*, April 15, 2019, https://www.latimes.com/opinion/editorials/la-ed-motor-voter-dmv-20190415-story.html.

68 registered the same day: Election Administration & Campaigns, Office of the Minnesota Secretary of State Steve Simon, https://www.sos.state.mn.us/election-administration-campaigns/data-maps/historical-voter-turnout-statistics.

70 mail ballots to all eligible voters: "Voting Outside the Polling Place: Absentee, All-Mail and Other Voting at Home Options," National Conference of State Legislatures, February 20, 2020, https://www.ncsl.org/research/elections-and-campaigns/absentee-and-early-voting.aspx.

70 "Every state could do this": Craig Timberg, "Voting By Mail, Already on the Rise, May Get a $500 Million Federal Boost from Coronavirus

Fears," *Washington Post*, March 10, 2020, https://www.washingtonpost.com/technology/2020/03/10/mail-voting-coronavirus-bill/.

70 who favor propose: Dale Ho, "Voting by Mail Will Save the 2020 Election," *New York Times*, March 12, 2020, https://www.nytimes.com/2020/03/12/opinion/coronavirus-election-vote-mail.html.

70 Officials must also consider: David Daley, "Coronavirus Could Normalize Voting by Mail. That Will Create Other Problems," *Washington Post*, March 12, 2020, https://www.washingtonpost.com/outlook/2020/03/12/coronavirus-vote-by-mail-problems/.

70 voter participation in vote-at-home states: Danielle Root and Liz Kennedy, "Increasing Voter Participation in America," Center for American Progress, July 11, 2018, https://www.americanprogress.org/issues/democracy/reports/2018/07/11/453319/increasing-voter-participation-america/.

70 voters ages eighteen to twenty-four outperformed: Gilad Edelman and Paul Glastris, "Letting People Vote at Home Increases Voter Turnout. Here's Proof," *Washington Post*, January 26, 2018, https://www.washingtonpost.com/outlook/letting-people-vote-at-home-increases-voter-turnout-heres-proof/2018/01/26/d637b9d2-017a-11e8-bb03-722769454f82_story.html.

Chapter Four: Transforming Non-Voters into Voters

74 save a few bigger dips, since 1974: "Youth Voting Historically," The Center for Information & Research on Civic Learning and Engagement (CIRCLE), https://civicyouth.org/quick-facts/youth-voting/.

74 a significant jump: "Voter Turnout Rates Among All Voting Age and Major Racial and Ethnic Groups Were Higher Than in 2014," Census.gov, https://www.census.gov/library/stories/2019/04/behind-2018-united-states-midterm-election-turnout.html.

75 Rock the Vote: "About Us," Rock the Vote, https://www.rockthevote.org/about-us/#ourhistory.

75 "we're at a turning point": Sheryl Crow, interview with author, October 20, 2018.

76 "parties and campaigns are having trouble": "Why Youth Don't Vote—Differences by Race and Education," CIRCLE, August 21, 2018, https://civicyouth.org/why-youth-dont-vote-differences-by-race-and-education/.

76 Newtown, Connecticut: "Harvard IOP youth poll finds stricter gun laws," Harvard Kennedy School Institute of Politics, June 18, 2018, https://iop.harvard.edu/about/newsletter-press-release/harvard-iop-youth-poll-finds-stricter-gun-laws-ban-assault-weapons.

76 "Because I vote all the time and you don't": Paul Gronke, interview with author, March 6, 2019.

76 "mobilized and cast a ballot": Hannah Mixdorf, interview with author, October 1, 2019.

78 concerned about climate change: We Vote Next Summit, "Voting Class of 2018," https://static1.squarespace.com/static/5a5fbce0d55b412c123047c3/t/5c070aba352f53d9bcd81889/1543965371297/2018+DELEGATE+YEARBOOK.pdf.

79 vote via a secure app: Terry Nguyen, "West Virginia to offer mobile blockchain voting app for overseas voters in November election," *Washington Post*, August 10, 2018, https://www.washingtonpost.com/technology/2018/08/10/west-virginia-pilots-mobile-blockchain-voting-app-overseas-voters-november-election/.

80 Inspire2Vote (previously known as Inspire U.S.): Hannah Mixdorf, Chelsea Costello, Olivia McCuskey of Inspire U.S. and Inspire2Vote (Inspire), interviews with author, January 23, June 26, and October 1, 2019.

80 transferred from student to student: Hanna Mixdorf, interview with author, October 2019; Laura Brill of The Civics Center, interview with author, October 1, 2019.

80 one-woman band: Andaya Sugayan, interview with author, September 27, 2019.

81 "To say that Barack threw himself into the job": Michelle Obama, *Becoming* (New York: Crown Publishing Group, 2018), 166–67.

82 A middle-aged teacher: Chelsea Costello (Inspire), interview with author, January 23, 2019.

82 voter registration drives in at least twenty-five states: Brill, interview with author, October 1, 2019.

82 "unboxing" video: The Civics Center, "Democracy in a box for HS Voter Registration Week," https://thecivicscenter.org/resources.

83 eighteen to nineteen, was only 23 percent nationwide: "Voter Turnout of Youth Aged 18–19 Shows States Having Varied Success at Growing Voters," CIRCLE, September 19, 2019, https://civicyouth.org/voter-turnout-of-youth-aged-18–19-shows-states-having-varied-success-at-growing-voters/.

83 who registered or pledged to vote through Inspire programs: Mixdorf (Inspire), interview with author, October 1, 2019.

83 less than 40 percent: Brill, interview with author, October 1, 2019; data available at Civics Center blog, https://thecivicscenter.org/blog.

86 wealthiest American households: Sean McElwee, "How Unequal Voter Turnout and Vote Suppression Helped Elect Donald Trump," *Salon*, May 14, 2017, https://www.salon.com/2017/05/14/how-unequal-voter-turnout-and-vote-suppression-helped-elect-donald-trump/.

86 The voters engaged by Nonprofit VOTE: Brian Miller, Caitlin Donnelly, and Caroline Mak, "Engaging New Voters," Nonprofit VOTE, May 1, 2019, https://www.nonprofitvote.org/documents/2019/05/engaging-new-voters-2018.pdf/.

87 Community Partnership Family Resource Center: Kathy Cefus, interview with author, September 25, 2019.

87 14 percent higher: Brian Miller et al., "Engaging New Voters," Nonprofit VOTE, May 1, 2019.

87 Nonprofit VOTE research and field coordinator: Caroline Mak, interview with author, June 19, 2019.

88 "When the people that we serve are voting": Jerome Sader, interview with author, September 13, 2019.

88 a hashtag that would encourage everyone: Kyle Lierman, interview with author, March 27, 2019.

89 "then get to the polls": Michelle Obama, "Create your VotingSquad," When We All Vote, October 26, 2018, https://action.whenweallvote.org/page/s/create-your-voting-squad.

89 "squad captains": *Voting Squad Starter Guide*, When We All Vote, https://www.whenweallvote.org/wpcontent/uploads/2019/12/WWAVVotingSquadGuide.pdf.

89 shortly after its launch: Jason Zengerle, "The Voices in Blue America's Head," *New York Times Magazine*, November 22, 2017, https://www.nytimes.com/2017/11/22/magazine/the-voices-in-blue-americas-head.html.

90 engaged listeners were also engaged voters: Shaniqua McClendon, interview with author, February 26, 2019.

91 "single best thing you can do": "Canvassing," *Pod Save America*, October 13, 2018, https://www.youtube.com/watch?v=GwghDw48iu0.

91 whichever candidate won the primary: Sridhar Pappu, "Trying to Flip the House, Zip Code by Zip Code," *New York Times*, July 20, 2018, https://www.nytimes.com/2018/07/20/business/swing-left-primary-campaigns.html.

91 why she'd volunteered: Adrienne Lever and Zoe Petrak, interview with author, November 1, 2018.

92 self-identified as evangelicals: Voter Education, Faith and Freedom Coalition, https://www.ffcoalition.com/voter-education/.

93 A table backstage: March for Our Lives members, interview with author, October 20, 2018.

94 Marjory Stoneman Douglas: Marjory Stoneman Douglas, National Women's Hall of Fame, https://www.womenofthehall.org/inductee/marjory-stoneman-douglas/.

Chapter Five: Making Voting Their Business

96 speaking with genuine concern about the North Dakota ID: Billy Ray Cyrus, interview with author, October 20, 2018.

96 and heightened brand awareness: Sofia Gross and Ashley Spillane, "Civic Responsibility: The Power of Companies to Increase Voter Turnout," Harvard Kennedy School Ash Center, June 2019, https://ash.harvard.edu/files/ash/files/harvard-casestudy-report-digital_copy.pdf.

96 bad news for the corporate world: Eric Orts, interview with author, May 23, 2019.

97 "culture shift around voting": Corley Kenna, interview with author, February 25, 2019.

97 "my way of taking action": Marissa Smith, interview with author, October 20, 2018.

98 head of Endeavor's government relations group: Amos Buhai, interviews with author, October 20, 2018, and February 27, 2019.

98 gave the rundown: Patrick Shanley, "Phone Banks, Paid Time Off and Parties: What Hollywood's Biggest Agencies Are Doing for Election Day," *Hollywood Reporter*, November 6, 2018, https://www.hollywoodreporter.com/lists/what-hollywoods-biggest-agencies-are-doing-election-day-1157082/item/caa-election-day-agencies-1158378.

98 mass shooting in Las Vegas: Rebecca Sun, "Hollywood's Stealth Support of the D.C. Gun-Control March: 'It's Important This Not Be the Celebrity Show,'" *Hollywood Reporter*, March 3, 2018, https://www.hollywoodreporter.com/news/hollywoods-stealth-support-dc-gun-control-march-important-not-be-celebrity-show-1095878.

100 company would close all of its offices: public blog post: Rose Marcario, Let Our People Go Vote, LinkedIn, June 14, 2018, https://www.linkedin.com/pulse/let-our-people-go-vote-rose-marcario/?trk=aff_src.aff-lilpar_c.partners_pkw.10078_net.mediapartner_plc.Skimbit%20Ltd._pcrid.449670_learning&veh=aff_src.aff-lilpar_c.partners_pkw.10078_net.mediapartner_plc.Skimbit%20Ltd._pcrid.449670_learning&irgwc=1.

102 Prosperity Candle: Kenna, interview with author, February 24, 2019.

102 their goal is 1,000: Kenna, email to author, January 6, 2020.

102 chief marketing officer: Adrianne Pasquarelli, "Levi's Promotes the Vote in New Campaign," *AdAge,* September 24, 2018.

104 after executives learned about the voting work: Kathy Cefus, interview with author, September 25, 2019.

104 On Election Day, the Iowa Association: Nicole Crain, interview with author, April 29, 2019.

104 more likely to vote: "2018 Employer to Employee Engagement Study," Business-Industry Political Action Committee, https://www.bipac.org/wp-content/uploads/2019/03/BIPAC_2018_E2E_Engagement_Study_031519.pdf.

105 "Stay in your lane": Hillary Kerr, interview with author, February 26, 2019.

106 got interested in politics and social action: Marissa Payne, "Group Led by Dolphins Owner Wants to See Every Professional Athlete Registered to Vote," *Washington Post*, September 24, 2017, https://www.washingtonpost.com/news/early-lead/wp/2017/09/24/group-led-by-dolphins-owner-wants-to-see-every-professional-athlete-registered-to-vote/.

107 twenty college athletic programs: Diahann Billings-Burford, Ian Cutler, Adam Wood (Rise to Vote), interview with author, May 13, 2019.

108 hosted a crowd: Billings-Burford, interview with author, May 13, 2019.

Chapter Six: Thank You for Voting

110 Alaska offered the animal stickers: Adelyn Baxter, "Vote Early to Get One of Juneau Artist Pat Race's 'I Voted' Stickers," Alaska Public Media, KTOO—Juneau, October 23, 2018, https://www.alaskapublic.org/2018/10/23/vote-early-to-get-one-of-juneau-artist-pat-races-i-voted-stickers/.

110 comparative Alaskan elections: Tegan Hanlon, "'It's Really Popular': Early Voting Numbers Are Up in Alaska Compared to Previous Elections for Governor," *Anchorage Daily News*, November 3, 2018.

110 how a particular factor affects an election: Paul Gronke, interview with author, March 6, 2019.

111 go-to book on the subject of encouraging voter turnout: Green and Gerber, *Get Out the Vote.*

111 robocalls: Green and Gerber, "Commercial Phone Banks, Volunteer Phone Banks, and Robocalls," in *Get Out the Vote*, chap. 6.

111 canvassing works: Green and Gerber, "Door-to-Door Canvassing," in *Get Out the Vote*, chap. 3.

111 Spanish-language soap opera: Green and Gerber, "Using Mass Media to Mobilize Voters," in *Get Out the Vote*, chap. 9.

111 social-pressure communications: Green and Gerber, "Strategies for Effective Messaging," in *Get Out the Vote*, chap. 11.

113 social-pressure-type messaging via text: Green and Gerber, "Electronic Mail, Social Media, and Text Messaging," in *Get Out the Vote*, chap. 7.

113 "preemptive thank-you note": Donald Green, interview with author, April 12, 2018.

113 people appreciate being appreciated: Green and Gerber, "Strategies for Effective Messaging."

114 tweaking language to ask people shortly before an election if they'll "be a voter": Christopher J. Bryan, Gregory M. Walton, Todd Rogers, and Carol S. Dweck, "Motivating Voter Turnout by Invoking the Self," *Proceedings of the National Academy of Sciences* 108, no. 31 (2011): 12653–12656, https://cpb-us-w2.wpmucdn.com/voices.uchicago.edu/dist/b/232/files/2016/09/Motivating-voter-turnout-by-invoking-the-self-1l8b75n.pdf.

114 fascinated by experiments: Christopher Bryan, interview with author, April 3, 2019.

115 study has prominent naysayers: Alan S. Gerber, Gregory A. Huber, Daniel R. Biggers, and David J. Hendry, "A Field Experiment Shows that Subtle Linguistic Cues Might Not Affect Voter Behavior," Proceedings

of the National Academy of Sciences, June 28, 2016, https://www.pnas.org/content/113/26/7112; Alan Gerber, Gregory Huber, and Albert Fang, "Do Subtle Linguistic Interventions Priming a Social Identity as a Voter Have Outsized Effects on Voter Turnout? Evidence from a New Replication Experiment: Outsized Turnout Effects of Subtle Linguistic Cues," *Political Psychology* 39, no. 4 (August 2018): 925–38.

115 follow-up paper: Christopher J. Bryan, David S. Yeager, and Joseph M. O'Brien, "Replicator Degrees of Freedom Allow Publication of Misleading 'Failures to Replicate,'" Proceedings of the National Academy of Sciences, 2019 (publication date pending).

115 inspired the "I am a voter" campaign: Mandana Dayani, interview with author, April 4, 2019.

117 "Sephora meets Coachella": Elizabeth Holmes, "Beauty Is in the Eye of These Beholders," *New York Times*, July 28, 2018.

117 working alongside them: Mandana Dayani, interview with author, May 29, 2019.

118 "very much a part of American electoral tradition": Donald Green, interview with author, April 12, 2018.

118 "raucous, freewheeling atmosphere": Green and Gerber, "Using Events to Draw Voters to the Polls," in *Get Out the Vote*, chap. 8.

119 Looking at nine festivals: Green and Gerber, "Using Events to Draw Voters to the Polls."

119 live musicians performing for hundreds: Shira Miller, interview with author, September 19, 2019.

120 "voter's illusion": Melissa Acevedo and Joachim I. Krueger, "Two Egocentric Sources of the Decision to Vote: The Voter's Illusion and the Belief in Personal Relevance," *Political Psychology* 25, no. 1 (2004).

121 never gained popularity among women in the United States in the same way: Katie Steinmetz, "Everything You Need to Know about the Word 'Suffragette,'" *Time*, October 22, 2015, https://time.com/4079176/suffragette-word-history-film/.

121 often colored for a particular candidate: ballot, "Did You Know?," Merriam-Webster, https://www.merriam-webster.com/dictionary/ballot#note-1.

121 "by papers": Jill Lepore, "Rock, Paper, Scissors," *The New Yorker*, October 6, 2008, https://www.newyorker.com/magazine/2008/10/13/rock-paper-scissors.

122 "party tickets": Lepore, "Rock, Paper, Scissors."

122 requiring the Australian ballot: Lepore, "Rock, Paper, Scissors."

122 Other words first used: Time Traveler, Merriam-Webster, https://www.merriam-webster.com/time-traveler/1944.

122 sociological gobbledygook": Philip Rocco, "Justice Roberts Said Political Science Is 'Sociological Gobbledygook.' Here's Why He Said It, and Why He's Mistaken," *Washington Post*, October 4, 2017, https://www

.washingtonpost.com/news/monkey-cage/wp/2017/10/04/justice-roberts-said-political-science-is-sociological-gobbledygook-heres-why-he-said-it-and-why-hes-mistaken/.

122 I Voted": Olivia B. Waxman, "This Is the Story Behind Your 'I Voted' Sticker," *Time*, November 6, 2018, https://time.com/4541760/i-voted-sticker-history-origins/.

122 Miami-Dade made sticker news: Douglas Hanks, "For Miami-Dade, a New 'I Voted' Sticker," *Miami Herald*, January 29, 2016, https://www.miamiherald.com/news/local/community/miami-dade/article57277743.html.

Chapter Seven: Gerrymandering: Over the Line?

128 expert in U.S. House of Representatives politics: David Wasserman, interview with author, May 16, 2019.

129 three out of thirteen congressional seats: Thomas Wolf and Peter Miller, "How Gerrymandering Kept Democrats from Winning Even More Seats Tuesday," *Washington Post*, November 8, 2018, https://www.washingtonpost.com/outlook/2018/11/08/how-gerrymandering-kept-democrats-winning-even-more-seats-tuesday/.

129 "hoverboard behind Peppa Pig": David Daley, *Ratf**ked* (New York: Liveright Publishing Corporation, 2017), 41.

129 "Goofy kicking Donald Duck": Mo Rocca, "Drawing the Lines on Gerrymandering," CBSNews.com, January 14, 2018, https://www.cbsnews.com/news/drawing-the-lines-on-gerrymandering/.

129 "fajita strip": Kevin Diaz, "Texas Gerrymandering Case before Supreme Court Could Change State's Political Map," *Houston Chronicle*, April 20, 2018, https://www.houstonchronicle.com/news/politics/texas/article/Gerrymandering-case-could-change-the-political-12851565.php#.

129 "broken-winged pterodactyl": Editorial Board, "Time for Maryland to Get Rid of 'Broken-winged Pterodactyl' Electoral Districts," *Washington Post*, January 27, 2016.

130 "all the marbles": Chris Jankowski, interview with author, May 20, 2019.

130 keeping cities or counties intact: Justin Levitt, "Where Are the Lines Drawn?" All About Redistricting, http://redistricting.lls.edu/where-state.php.

130 electing a candidate of their choice: *Thornburg v. Gingles*, U.S. Supreme Court, 478 U.S. 30 (1986), https://supreme.justia.com/cases/federal/us/478/30/#tab-opinion-1956757.

131 fewer Democratic districts: Galen Druke, "Is Gerrymandering the Best Way to Make Sure Black Voters Are Represented?," FiveThirtyEight, December 14, 2017, https://fivethirtyeight.com/features/is-gerrymandering-the-best-way-to-make-sure-black-voters-are-represented/.

131 require state legislatures: Brennan Center for Justice, "Who Draws the Maps?

Legislative and Congressional Redistricting," January 30, 2019, https://www.brennancenter.org/our-work/research-reports/who-draws-maps-legislative-and-congressional-redistricting.

132 can predict voter behavior: Vann R. Newkirk II, "How Redistricting Became a Technological Arms Race," *The Atlantic,* October 28, 2017, https:/www.theatlantic.com/politics/archive/2017/10/gerrymandering-technology-redmap-2020/543888/.

132 "generated five": Daley, *Ratf**ked*, xxiv.

132 poured money into: Daley, *Ratf**ked*, xix.

132 Jankowski was one: Chris Jankowski, "How a Political Consultant Changed Voting Districts Nationally," interview by Karen Duffin, Planet Money, NPR, June 15, 2018, https://www.npr.org/2018/06/15/620471099/how-a-political-consultant-changed-voting-districts-nationally.

133 unconstitutional racial gerrymanders: *Cooper v. Harris*, U.S. Supreme Court, 581 US __ (2017), https://supreme.justia.com/cases/federal/us/581/15-1262/.

133 "to create as many": *Rucho v. Common Cause*, U.S. Supreme Court, Case No. 18–422, Brief for Appellees, League of Women Voters of North Carolina, 6, https://campaignlegal.org/sites/default/files/2019-03/18-422%20Brief%20of%20Appellee_FINAL%20PDF-A.pdf.

133 after his death: Michael Wines, "The Battle Over the Files of a Gerrymandering Mastermind," *New York Times*, September 4, 2019, https://www.nytimes.com/2019/09/04/us/gerrymander-north-carolina-hofeller.html.

135 fairly drawn district: *Rucho v. Common Cause*, U.S. Supreme Court, Case No. 18–422, Brief for Appellees, League of Women Voters of North Carolina, 1.

135 "inherently political task": *Rucho v. Common Cause*, U.S. Supreme Court, Case No. 18–422, Brief for Appellants, 7, https://www.supremecourt.gov/DocketPDF/18/18-422/92363/20190319113346714_18-422%20rb.pdf.

135 how much partisanship is too much: *Rucho v. Common Cause*, U.S. Supreme Court, Case No. 18–422, 2019, Brief for Appellants, 11.

135 justices punted: Adam Liptak, "Supreme Court Avoids an Answer on Partisan Gerrymandering," *New York Times*, June 18, 2018, https://www.nytimes.com/2018/06/18/us/politics/supreme-court-wisconsin-maryland-gerrymander-vote.html.

137 Democratic leaders had managed to gerrymander: *Lamone v. Benisek*, U.S. Supreme Court, Case No. 18–726 (2019), https://www.oyez.org/cases/2018/18-726.

137 "change their behavior": *Lamone v. Benisek*, U.S. Supreme Court, oral argument transcript, Case No. 18–726 (2019), March 26, 2019, 11, https://www.supremecourt.gov/oral_arguments/argument_transcripts/2018/18-726_9olb.pdf.

137 "not going to dispute that": *Rucho v. Common Cause*, U.S. Supreme Court, oral argument transcript, Case No. 18–422 (2019), March 26, 2019, 68, https://www.supremecourt.gov/oral_arguments/argument_transcripts/2018/18-422_5hd5.pdf.

137 *Rucho v. Common Cause*, U.S. Supreme Court, 588 U.S. ___ (2019), June 27, 2019, opinion, https://supreme.justia.com/cases/federal/us/588/18–422/#tab-opinion-4114540; dissent, https://supreme.justia.com/cases/federal/us/588/18–422/#tab-opinion-4114540.

138 citizen redistricting commission: Katie Fleming, "Voters Pass Redistricting Reform in California, Florida, and Minnesota," Common Cause California, February 27, 2014, https://www.commoncause.org/california/press-release/voters-pass-redistricting-reforms-in-california-florida-and-minnesota/.

139 Riggs said sharply: *Rucho v. Common Cause*, U.S. Supreme Court, oral argument transcript, Case No. 18–422 (2019), March 26, 2019, 69–71.

139 In a bipartisan poll: New Bipartisan Poll on Gerrymandering and the Supreme Court (memorandum), January 25, 2019, https://campaignlegal.org/sites/default/files/2019–01/CLC%20Bipartisan%20Redistrictig%20Poll.pdf.

139 "cautiously optimistic": Paul Smith, interview with author, March 26, 2019.

140 will of the people: Felicia Sonmez and Robert Barnes, "North Carolina Court Rules Partisan State Legislative Districts Unconstitutional," *Washington Post*, September 3, 2019, https://www.washingtonpost.com/politics/north-carolina-court-rules-partisan-state-legislative-districts-unconstitutional/2019/09/03/4a137034-ce8a-11e9-8c1c-7c8ee785b855_story.html.

140 "sparked the popular uprising": Paul Smith, email to author, June 30, 2019.

140 had to be a way: Katie Fahey, interview with author, April 25, 2019.

141 went all the way: Jonathan Oosting, "Mich. Supreme Court: Redistricting Plan Goes on Nov. Ballot," *Detroit News*, August 1, 2018, https://www.detroitnews.com/story/news/local/michigan/2018/07/31/michigan-supreme-court-gerrymandering-initiative/871624002/.

141 "had gerrymandered the state": Fahey, interview with author, April 25, 2019.

142 "founding of our nation": Fahey, interview with author, April 25, 2019.

144 "key political films": Owen Gleiberman, "Tribeca Film Review: 'Slay the Dragon,'" *Variety*, April 28, 2019, https://variety.com/2019/film/reviews/slay-the-dragon-review-gerrymandering-1203199856/.

144 "bring it to life": Chris Durrance and Barak Goodman, interview with author, August 29, 2019.

144 "a rallying cry": Neal Block, interview with author, September 12, 2019.

Chapter Eight: Knowing the News Is Real

146 colonial anecdote illustrating the consequences: Jackie Mansky, "The Age-Old Problem of Fake News," Smithsonian.com, May 7, 2018, https://www.smithsonianmag.com/history/age-old-problem-fake-news-180968945/.

146 learned something: Jeffrey Gottfried, Michael Barthel, Elisa Shearer, and Amy Mitchell, "The 2016 Presidential Campaign—a News Event That's Hard to Miss," Pew Research Center, February 4, 2016, https://www.journalism.org/2016/02/04/the-2016-presidential-campaign-a-news-event-thats-hard-to-miss/.

146 "exhausted by the amount of election coverage": Jeffrey Gottfried, "Most Americans Already Feel Election Coverage Fatigue," Pew Research Center, July 14, 2016, https://www.pewresearch.org/fact-tank/2016/07/14/most-americans-already-feel-election-coverage-fatigue/.

147 distinguish factual news statements: Amy Mitchell, Jeffrey Gottfried, Michael Barthel, and Nami Sumida, "Distinguishing Between Factual and Opinion Statements in the News," Pew Research Center, June 18, 2018, https://www.journalism.org/2018/06/18/distinguishing-between-factual-and-opinion-statements-in-the-news/.

147 "incredible information literally available at our fingertips": Alan Miller, interview with author, April 16, 2019.

147 "intoxicating casino": Brian Stelter, interview with author, April 24, 2019.

148 complain about coverage: Mansky, "The Age-old Problem of Fake News," Smithsonian.com, May 7, 2018.

154 "standards, biases, or beliefs": Michael Schmidt, interview with author, November 26, 2018.

154 false information has legs: Craig Silverman et al., "Hyperpartisan Facebook Pages Are Publishing False and Misleading Information at an Alarming Rate," *BuzzFeed*, October 20, 2016, https://www.buzzfeednews.com/article/craigsilverman/partisan-fb-pages-analysis#.tom4Bwyro.

154 breakdown of what various titles mean: Brian Stelter, interview with author, April 24, 2019.

154 ballot initiatives as deserving: Jeanette Senecal, interview with author, September 18, 2019.

154 notoriously long: Corley Kenna, interview with author, February 25, 2019.

154 call about a particular issue: Hillary Kerr, interview with author, February 26, 2019.

Chapter Nine: Understanding Polling

164 win each state, was hedging: "Elections Podcast Countdown, Final Election Preview," FiveThirtyEight, November 6, 2018, https://fivethirtyeight.com/features/elections-podcast-countdown-final-election-preview/.

164 the ad will help the candidate: Philip Bump, interview with author, June 16, 2019.

165 "extremely" interested: Dana Blanton, "Fox News Poll: Interest in 2020 Already at Election Day Levels," FoxNews.com, April 18, 2019, https://www.foxnews.com/politics/fox-news-poll-interest-in-2020-already-at-election-day-levels.

165 comprehensive evaluation: "An Evaluation of 2016 Election Polls in the U.S.," Ad Hoc Committee on 2016 Election Polling, American Association for Public Opinion Research, May 4, 2017, https://www.aapor.org/Education-Resources/Reports/An-Evaluation-of-2016-Election-Polls-in-the-U-S.aspx.

167 what went wrong in 2016: "Next Question with Katie Couric," interview with Clare Malone, November 1, 2018, https://www.stitcher.com/podcast/how-stuff-works/katie-couric/e/56999249?autoplay=true.

168 helps shape understanding: Kristen Soltis Anderson and Margie Omero, interview with author, November 26, 2018.

168 more weight to polls: "A User's Guide to FiveThirtyEight's General Election Forecast," FiveThirtyEight, June 29, 2016, https://fivethirtyeight.com/features/a-users-guide-to-fivethirtyeights-2016-general-election-forecast/.

171 an expert on public polling: Karlyn Bowman, interview with author, May 2, 2019.

171 "probabilistic forecasts can give potential voters": Solomon Messing, "Use of Election Forecasts in Campaign Coverage Can Confuse Voters and May Lower Turnout," Pew Research Center, February 6, 2018, https://www.pewresearch.org/fact-tank/2018/02/06/use-of-election-forecasts-in-campaign-coverage-can-confuse-voters-and-may-lower-turnout/.

171 "Margin of error" is a technical-sounding phrase: Andrew Mercer, "5 Key Things to Know About the Margin of Error in Election Polls," Pew Research Center, September 6, 2016, https://www.pewresearch.org/fact-tank/2016/09/08/understanding-the-margin-of-error-in-election-polls/.

171 glance through the history of the Gallup Poll: Gregor Aisch and Alicia Parlapiano, "What Do You Think Is the Most Important Problem Facing This Country Today?," *New York Times*, February 27, 2017, https://www.nytimes.com/interactive/2017/02/27/us/politics/most-important-problem-gallup-polling-question.html.

171 "government/poor leadership": "Most Important Problem," Gallup, https://news.gallup.com/poll/1675/most-important-problem.aspx.

Chapter Ten: Explaining the Electoral College

175 Republican elector Art Sisneros: Michael Marks, "This Texas Elector Resigned Rather Than Voting for Donald Trump," KUT.org, November 29,

2016, https://www.kut.org/post/texas-elector-resigned-rather-voting-donald-trump.

176 78 percent of the vote in Liberty County: "2016 Texas Presidential Election Results," Politico.com, December 31, 2016, https://www.politico.com/2016-election/results/map/president/texas/.

176 there was silence: Patrick Svitek, Bobby Blanchard, and Aliyya Swaby, "Texas Electors Cast 36 Votes for Trump, 1 for Kasich and 1 for Ron Paul," *Texas Tribune* (video available), December 19, 2016, https://www.texastribune.org/2016/12/19/watch-texas-electoral-college-vote-begins-texas-ca/.

178 "to be delegates rather than trustees": Sandford Levinson, interview with author, May 8, 2019.

179 bound their electors: "Summary: State Laws Regarding Presidential Electors," National Association of Secretaries of State, November 2016, https://www.nass.org/sites/default/files/surveys/2017–08/research-state-laws-pres-electors-nov16.pdf.

180 calling into question: Meagan Flynn, "He Tried to Stop Trump in the Electoral College. A Court Says His 'Faithless' Ballot Was Legal," *Washington Post*, August 22, 2019, https://www.washingtonpost.com/nation/2019/08/22/he-tried-stop-trump-electoral-college-court-says-his-faithless-ballot-was-legal/.

180 chosen for their loyalty: John Fortier, email to author, August 22, 2019.

181 unique among democracies: Drew DeSilver, "Among Democracies, U.S. Stands Out in How It Chooses Its Head of State," Pew Research Center, November 22, 2016, https://www.pewresearch.org/fact-tank/2016/11/22/among-democracies-u-s-stands-out-in-how-it-chooses-its-head-of-state/.

182 finding a compelling reason: Alexander Keyssar, interview with author, May 21, 2019; Garrett Epps, "The Electoral College Was There from the Start," *The Atlantic*, September 8, 2019, https://www.theatlantic.com/ideas/archive/2019/09/electoral-college-terrible/597589/.

182 pacify framers from slaveholding states: Sean Wilentz, "The Electoral College Was Not a Pro-Slavery Ploy," *New York Times*, April 4, 2019, https://www.nytimes.com/2019/04/04/opinion/the-electoral-college-slavery myth.html; Akhil Reed Amar, "Actually, the Electoral College Was a Pro-Slavery Ploy," *New York Times*, April 6, 2019, https://www.nytimes.com/2019/04/06/opinion/electoral-college-slavery.html.

182 "particular weird institution": Alexander Keyssar, interview with author, May 21, 2019.

184 "role of kingmaker": Norman J. Ornstein, "Three Disputed Elections: 1800, 1824, 1876," in *After the People Vote: A Guide to the Electoral College*, ed. John C. Fortier, 3d ed. (Washington, DC: The AEI Press, 2004), chap. 7.

185 seven hundred proposed bills or amendments: "Past Attempts at Reform," FairVote, https://www.fairvote.org/past_attempts_at_reform.

185 generally unpopular: Jeffrey M. Jones, "American Split on Proposals for Popular Vote," Gallup, May 14, 2019, https://news.gallup.com/poll/257594/americans-split-proposals-popular-vote.aspx.

185 gained Republican support: Art Smith, "Americans' Support for Electoral College Rises Sharply," Gallup, December 2, 2016, https://news.gallup.com/poll/198917/americans-support-electoral-college-rises-sharply.aspx.

186 House showed no such enthusiasm: "Past Attempts at Reform," FairVote, https://www.fairvote.org/past_attempts_at_reform.

186 Despite its popularity: Lily Rothman, "The Electoral College Votes Today. But Politicians Have Been Trying to Reform It for Decades," *Time*, December 19, 2016, https://time.com/4597833/electoral-college-donald-trump-challenge/.

187 said that Hayes had stolen from dead soldiers and shot his own mother: Ornstein, "Three Disputed Elections: 1800, 1824, 1876."

187 believed until his dying day: Ornstein, "Three Disputed Elections: 1800, 1824, 1876."

188 population numbers don't warrant: Levinson, interview with author, May 8, 2019.

188 own supposed detriment: Alexander Keyssar, interview with author, May 21, 2019.

189 combined 196 electoral votes: Caroline Kelly, "Oregon Governor Signs Bill Granting State's Electoral Votes to National Popular Vote Winner," CNN.com, June 12, 2019, https://www.cnn.com/2019/06/12/politics/oregon-joins-national-popular-vote-compact/index.html.

190 untested system: John Fortier, interview with author, April 24, 2019.

INDEX

abolition, of slavery, 29
Abrams, Stacey, 60, 90, 143
absentee ballot, state election laws on, 47
ACLU. *See* American Civil Liberties Union
Adams, Abigail, 25–26
Adams, John, 25–26, 180, 185
  on citizenship demand, 6
  on property ownership, 5
African Americans
  Civil War and votes of, 7, 29
  discrimination of, 7
  disenfranchisement of, 10
  early days voting prohibition, 6
  Equal Rights League, 8
  felon voting and, 49
  Fourteenth Amendment, 8
  Grandfather Clause and, 10
  Ku Klux Klan and, 9
  political offices of, 8–9
  Reconstruction and, 8–9
  voting of, 5–16
  voting prohibition, in early days for, 6
  voting rights for, x, 43
African American women
  Nineteenth Amendment and, 43
  in parades and protests, 37–38
  support of rights of, 30–32
  VRA and, 43
  in women's suffrage, 30–31, 34–35
After the People Vote (Ornstein and Fortier), 184, 189–90
"Ain't I a Woman" (Truth), 28
All On The Line, 143
American Association for Public Opinion Research, 165
American Civil Liberties Union (ACLU)
  on felon voting, 49
  on voter training requirements and registration, 48
American Enterprise Institute, 168
American Equal Rights Association, 30
American Woman Suffrage Association, 30
analyst, on cable news panel, 159
anchor, on cable news panel, 159
Anderson, Kristen Soltis, 167–68, 172
Anthony, Susan B., 24, 28, 40, 163
  obituary, 43
  voting fine for, 32
apathy, youth vote and, 93
"Appeal of Forty Thousand Citizens, Threatened with Disfranchisement" (Purvis), 6
Asian Americans
  presidential election voting of, xii
  voting rights for, x
Associated Press, 150–51
automatic voter registration (AVR), 99
  benefits of, 66
  Bird on, 69
  Brennan Center study on, 67, 69
  DMV interaction for, 66, 68
  FiveThirtyEight study on, 68
  Gronke on, 66–67
  registration increase from, 67–68
  2016 presidential election and, 67

ballot, 121
  absentee, 47
  news media and, 147–48
Ballotpedia, 195
Becoming (Obama, M.), 81
Berman, Ari, 14, 52–53

Billings-Burford, Diahann, 108
Bill of Rights, free press in, 148–50
BIPAC. *See* Business-Industry Political Action Committee
Bird, Jeremy, 46–47, 69
Blatch, Harriot Stanton, 35–36, 41
Bloody Sunday, VRA and, 13–14
Blue Cross Blue Shield, voting initiative of, 101
book sources, for news media trust, 152
Bowman, Karlyn, 168–70, 172–73
Boxer, Barbara, 184
Brennan Center
  AVR study by, 67, 69
  voter rolls purging study by, 59
Breyer, Stephen, 137
Britain, women's suffrage in, 35–36
Brown v. Board of Education (1954), 13
Bryan, Christopher, 114–15
Bryce, James, 47
Buhai, Arnos, 98–99
Bump, Philip, 164
Burr, Aaron, 180
Bush, George W., 122, 166, 185, 188
Business-Industry Political Action Committee (BIPAC), 104–5
business voting initiatives
  BIPAC on, 104–5
  of Blue Cross Blue Shield, 101
  of Clique Brands, 105–6
  employer-led, 107–8
  of Endeavor Part at the Polls, 97–99
  of Iowa Association of Business and Industry, 104
  on-site employers, 103–5
  of Patagonia Time to Vote, 95, 97, 100–103
  personal voting awakenings, 105–6
  of RISE, 107–8
  voting encouragement steps for, 102–3
Buzzfeed analysis, on false news information, 157–58
cable news panel
  analyst on, 159
  anchor on, 159
  contributor or commentator on, 159
  correspondent on, 158–59
Campaign Legal Center, on gerrymandering, 139
candidate platforms, as voting motivation, xiii
canvassing, 91–92, 111
  door-to-door, 74
capitalism, democracy support of, 96
Carter, Jimmy, 53–54
Catt, Carrie Chapman, 24–25
celebrity power, for peer pressure, 88–89
chad, 122, 127
checklist, for voting, 193–97
Chinese Exclusion Act (1882), 20–21
CIRCLE, on youth vote, 75–76
citizenship
  Adams, J., on demand for, 6
  Fourteenth Amendment on, 8
  of Native Americans, 16–17
civic duty
  voting as, 101–2
  as voting motivation, xii
Civic Nation, 118–19
The Civics Center, peer-to-peer registration by, 80, 82–83
Civil War
  African Americans vote and, 7, 29
  felons voting after, 49
Clay, Henry, 184
Cleveland, Grover, 185
climate strike, *45*
Clinton, Bill, 75, 179
Clinton, Hillary
  popular vote of, 179–80, 184
  supporters of, *163*
  2016 election and, 163, 183–84, 188
Clique Brands, voting initiative of, 105–6
Common Cause v. Lewis (2019), 139–40

Constitution, U.S.
Electoral College amendment and, 185–86
First Amendment, 139, 148–50
Twelfth Amendment, 180
Fourteenth Amendment, 8, 29, 32
Fifteenth Amendment, 6, 9, 15, 16, 30, 32
Nineteenth Amendment, 6, 32, 40–43
Twenty-Sixth Amendment, 22
Contreras, Ramon, 93
contributor or commentator, on cable news panel, 159
Cooper, Anderson, 159
coronavirus, xv, 70, 89
correspondent, on cable news panel, 158–59
Couric, Katie, 167
Crain, Nicole, 104–5
Crooked Media, 89
McClendon of, 90
Crow, Sheryl, 75, 93, 95–96
Curry, Steph, 115, 117
Cyrus, Billy Ray, 95–96

Daley, David, xiv
on partisan gerrymandering, 132–34
Dawes Act (1887), 16–17
Dayani, Mandana, 115–17
Deitsch, Matt, 93
democracy
capitalism supported by, 96
free enterprise system and, 96–97
Department of Motor Vehicles (DMV), AVR and, 66, 68
discrimination
of African Americans, 7
against Native Americans, 16
women's suffrage and, 29–32
disenfranchisement
of African Americans, 10
of women, 29
diversification, of news media, 157–58
DMV. *See* Department of Motor Vehicles
Dodson, Angela P., 31
door-to-door canvassing, 74
Douglass, Frederick, 28, 30, 31
draft age, youth vote impact by, 21
Dred Scott case, 7, 13
Duff, Brendan, 93
Duff, Daniel, 93
Dunlap, Matt, 64
Durrance, Chris, 142

early voting, state election laws on, 47
education, xii
women vote on, 26
Eighteen x 18, 115
Election Day
early days of voting and, 4
in modern world, 4
state election law on voter ID for, 57
youth vote on, 83
elections. *See also* 2016 presidential election
of 1800, 180
of 1824, 184
of 1876, 186–87
gerrymandering and 2020, 143–44
skipping of and voter rolls purging, 59–60
2012 Obama, B., reelection campaign, 46
voter rolls purging and notice response failure, 59–60
Electoral College, xiii
Boxer on, 184
in election of 1800, 180
in election of 1824, 184
in election of 1876, 186–87
as entertainment, 177–78
FairVote on, 185
Hamilton on, 181–82
history of, 181–84
Keyssar on, 188
Levinson on, 188

Electoral College (*continued*)
National Popular Vote Interstate Compact and, 189–90
popular vote and, xiv, 183, 185, 190–91
proposals to end, 184–86
protest on, 175
slavery impact on history of, 182
small states and, 187–89
state population and, 179
swing states, 187–89
Trump 2016 election and, 166, 177, 184, 188
electors, in Electoral College
faithless electors, 177
Levinson on, 178
as mirrors, not trustees, 178–80
Paul, R., and Kasich as, 177
political party loyalty and, 179–80
Sisneros as, 175–77
states choice of, 179
Supreme Court on removal of, 180
employees
Endeavor registration goals for, 99
Patagonia support of vote of, 95, 97, 100–103
employer-led, business voting initiatives of, 107–8
Endeavor Part at the Polls
employee registration goals, 99
on gun reform, 98
Smith, M., as organizer of, 97–98
voting promotion efforts by, 98–99
Equal Rights League, 8
exclusion
Keyssar on, 50
Remnick on, 51

FactCheck.org, 151
Fahey, Katie, 140–43
Fair Fight 2020, 60, 90, 143
fair map advocates, on gerrymandering, 138
FairVote advocacy group
on Electoral College, 185
REDMAP of, 132–33
Faith and Freedom Coalition, 92
faithless electors, 177
false news information, *Buzzfeed* analysis on, 157–58
Favreau, Jon, 89–90
federal elections, voting age reduction for, 22
felons, voting by
ACLU on, 49
African Americans, 49
state election laws on, 48–50
unconstitutional poll tax and, 50
Fifteenth Amendment, 32
men of all races voting rights, 6, 9, 30
Native Americans and, 16
VRA and, 15
First Amendment
free press and, 148–50
voter ID and, 139
FiveThirtyEight, 163–64, 167
AVR study by, 68
on polling, 168
on voter ID laws, 58
Fortier, John, 184, 189–90
Fourteenth Amendment, 8, 29, 32
free black men, voting by, 5–6
free enterprise system, 96–97
free press, 148–50

Gallup Poll, 171, 172
Gerber, Alan S., 111–12
on get-out-the-vote language methods, 115
Gerlach, Jim, 104–5
Gerry, Elbridge, 127–28
gerrymandering, xiii
Campaign Legal Center on, 139
Fahey "Slay the Dragon" campaign on, 140–43
fair map advocates, 138

future on, 138–40
Hofeller as master of, 133
Holder on, 136
Jankowski on, 130
origin of, 127–28
partisan, 130–38
racial, 130–31, 133, 135
rally to end, *127*
Roberts, J., on, 122
Schwarzenegger and, 136
state election laws on, 138–39
Supreme Court on partisan, 134–38
2020 election and, 143–44
Wasserman on, 128
get-out-the-vote efforts, x, 74–75
Endeavor nonpartisan, 98–99
organizations for, 63
Get Out the Vote: How to Increase Voter Turnout (Green and Gerber), 111–12
get-out-the-vote language methods, 116–17
Bryan research on, 114–15
of Eighteen x 18, 115
Gerber on, 115
of Inspire2Vote, 115
of When We All Vote, 115
get-out-the-vote movements, xiii, 93
Ginsburg, Ruth Bader, 83
RBG documentary on, 142
Give Us the Ballot (Berman), 14, 52–53
gobbledygook, 122
Goodman, Barak, 142
Gore, Al, 122, 185, 188
Gorsuch, Neil, 139
Grandfather Clause, 10
Green, Donald
on motivation for voting, xv
on voter encouragement parties, 118–19
on voter encouragement tactics, 111–12
Gronke, Paul, 76
on AVR, 66–67
on voting factors, 110–11
guardianship, of Native Americans, 17
gun reform
Endeavor on, 98
youth vote for, 76
Gurung, Prabal, 115

Hamilton (Miranda), 4
Hamilton, Alexander, 180
on Electoral College, 181–82
Harlan, John Marshall, 13
Harrison, Benjamin, 10–11, 185
Hayes, Rutherford B., 9–10, 185, 186–87
high school population, registration of, 76, 80
Ho, Dale, 49, 62
on voter suppression, 56–57
Hofeller, Thomas, 133
Holder, Eric
All On The Line and, 143
on gerrymandering, 136
Hopkins, Dan, 58
Housing Action Illinois, 88
Hughes, Akilah, 91
Hunt, Jane, 27
Huseman, Jessica, 62
Husted v. A. Philip Randolph Institute (2018), 59
Hutchinson, Thomas, 145–46
Hyde, Henry, 54

"I am a voter" promotions, 109–23
Dayani creation of, 115–17
get-out-the-vote language methods, 114–17
#ididitforthesticker, 109
"I Voted" stickers, 109–10, 122–23, 163
parties to encourage, 118–19
voter encouragement tactics, 102–3, 111–13
voter illusions, 119–20
voting language facts, 120–23

"I am a voter" sign, Traugott and, *109*
immigrants
  Know-Nothings group and, 20
  literacy tests and, 20
  2016 presidential election vote and, 116
  vote of, 19–21
  voter suppression of, 53
  VRA renewal and, 53
inclusion, Keyssar on, 50
inconsistent responders, for 2016 presidential election, 166
Indian Citizenship Act. *See* Snyder Act
Inskeep, Steve, 43
Inspire2Vote, 76, 81
  get-out-the-vote language of, 115
  peer-to-peer registration by, 80, 83
interactive voice response (IVR), in polling, 171
international voting requirements, 46
Iowa Association of Business and Industry, voting initiative of, 104
Isbell, Jason, 96
"I Voted" stickers, 109–10, 122–23, 163
IVR. *See* interactive voice response

Jackson, Andrew, 184, 185
Jackson, Jimmie Lee, 14
Jankowski, Chris, 130, 132–33, 143–44
Jefferson, Thomas, 180
Jim Crow laws, 10
  background of, 11–12
  on literacy tests, 11, 13
  on poll taxes, 11, 13
Johnson, Lyndon, 14

Kagan, Elena, 136–38
Kaiser Family Foundation, 171
Kasich, John, 177, 179–80
Kavanaugh, Brett, 136, 137–38
Kenna, Corley, 97, 161
  on Patagonia voting promotion, 100–101
Kennedy, Anthony, 135–36
Kennedy, Edward, 22
Kerr, Hillary, 105–6, 161
Keyssar, Alexander, 7, 9
  on Electoral College, 188
  on inclusion and exclusion, 50
  on political elite, 20
King, John, 4
King, Martin Luther, Jr., 53
Know-Nothings group, immigrants and, 20
Kobach, Kris, 62
Ku Klux Klan, 9

language, of voting
  ballot, 121
  chad, 122, 127
  facts of, 120–23
  get-out-the-vote methods, 114–17
  gobbledygook, 122
  "I Voted" sticker, 109–10, 122–23, 163
  party ticket, 121–22
  for polling, 167
  social-pressure communications, 111–12
  suffragette, 120–21
  for youth vote, 77–79, 115–16
Lawrence v. Texas (2018), x, 139
League of Women Voters, *ix*
  People Powered Fair Maps of, 143
  on racial gerrymandering, 135
  VOTE411 of, 63, 148, 160, 193, 195
  on voter rolls purging, 62–63
  voter suppression fight by, 58–59
learned behavior, for youth vote, 84–86
Lepore, Jill, 121
Lever, Adrienne, 92
Levinson, Sanford, 178, 188
Lewis, John, 14
Lierman, Kyle, 88
Liptak, Adam, 55
literacy tests
  immigrants and, 20
  Jim Crow laws on, 11, 13
litigation, on partisan gerrymandering, 132–33
Lodge, Henry Cabot, 10–11

Lott, Eric, 11–12
Lovett, Jon, 89, 91
low-income Americans, voter turnout of, 86

magazine sources, for news media trust, 151–52
Magnuson Act (1943), 20–21
majority-minority districts, racial gerrymandering and, 130
Mak, Caroline, 87
Malone, Clare, 167
Marcario, Rose, 100
March for Our Lives
 get-out-the movement of, xiii, 93
 performing artists at, 95–96
margin of error, polling and, 170
Marziani, Mimi, 51
McClendon, Shaniqua, 90
McConnell, Kirsten, 94
McCool, Daniel, 18–19
 as Native American rights advocate, 55
Messing, Solomon, 169
Mexican American Legislative Caucus, 56
Milholland, Inez, 36
Miller, Alan, 147
Miller, Shira, 119
minority groups, Voting Rights Act on vote of, 6–7
Miranda, Lin-Manuel, 4
Mixdorf, Hannah, 76, 81
monitoring, of voter rolls, 58–61
motivations, for voting
 candidate platforms, xiii
 civic duty, xii
 Green on, xv
 political party loyalty, xii
 union as, xii
Mott, Lucretia, 27
movements
 get-out-the-vote, xiii, 93
 Rock the Vote youth, 75
 for women's suffrage, 27–29
 for youth vote, 75–76
National American Woman Suffrage Association (NAWSA), 35, 37–38
National Association of Colored Women, 30–31
National Basketball Association (NBA), 108
National Council of Women, 31
National Education Association, 21
National Football League (NFL), 107
Nationality Act (1952), 21
national networks, for news media trust, 150
National Popular Vote Interstate Compact, 189–90
Native Americans
 citizenship and, 16–17
 discrimination against, 16
 Fifteenth Amendment and, 16
 guardianship of, 17
 McCool as rights advocate for, 55
 poll tax of, 17–18
 state election laws on vote of, 17
 vote of, 16–19
 voter ID law and, 51, 96
 voting rights for, x
 WWII veterans and, 17
Native Vote (McCool), 18–19
Naturalization Act (1790), 19
NAWSA. *See* National American Woman Suffrage Association
NBA. *See* National Basketball Association
News Literacy Project, of Miller, A., 147
news media, 145. *See also* trust, of news media
 ballot and, 147–48
 cable news panel, 158–59
 diversification of, 157–58
 free press role in, 148–50
 non-news sources, 160–62
 political awareness and, 147
 proper news story and, 153–55
 Shahidi on, 146

news media (*continued*)
unbiased sources in, 149
unfair story and, 155–57
The New Yorker, 151–52
NFL. *See* National Football League
Nineteenth Amendment, 32, 40–42
African American women and, 43
women suffrage and, 6
Nixon, Richard, 53
non-news sources, 160, 162
Kenna as, 161
Kerr as, 161
Nonprofit VOTE, 86–88
numbers, polling and, 168–69

Obama, Barack, 73–74, 132
All On The Line and, 143
2012 reelection campaign, 46
wave election and, 133
Obama, Michelle, 81
When We All Vote of, 88–91, 115
online polling, 171
on-site employers, business voting initiatives of, 103–5
Ornstein, Norman, 184, 189–90
Orts, Eric, 96–97

parades and protests
African American women in, 37–38
Blatch on women, 36
on Electoral College, *175*
Paul, A., in, 37
Roberts, R., on, 37
for women's suffrage, 35–38
parties, for voter encouragement
American electoral tradition of, 118
Civic Nation on, 118–19
constitutionality of, 135
Green on, 118–19
#VoteTogether, 118, 119, 196
partisan gerrymandering, 130, 131
Daley on, 132–34
Kavanaugh on, 136, 137–38
Kennedy, A., on, 135–36
litigation on, 132–33
Supreme Court on, 134–38
party ticket, 121–22
Patagonia offices, *95*
employees voting support by, 97, 100–103
Paul, Alice, 37
Paul, Ron, 177
peer pressure, 88–89
peer-to-peer registration, 80, 82–83
Pelosi, Nancy, 23–24
People Powered Fair Maps, of League of Women Voters, 143
Pérez, Myrna, 69
performing artists, at March for Our Lives, 95–96
personal importance, for youth vote, 93–94
Petrak, Zoe, 91–92
Pew Research Poll, 171
Plessy v. Ferguson (1896), 12
Harlan on, 13
podcasts
for news media trust, 152
Pod Save America on, 73–74, 89–91
The Pollsters, 167–68, 172
Pod Save America HBO special, 73–74, 89–91
political awareness, news media and, 147
political elite, Keyssar on, 20
political equality, Roosevelt on, xi
political offices
of African Americans, 8–9
of women, 24
political parties
electors loyalty to, 179–80
loyalty, as voting motivation, xii
women's suffrage and, 33–34
Politico, 151
polling, xiii
American Association for Public Opinion Research on, 165

Bowman on, 168–70, 172–73
FiveThirtyEight on, 168
IVR in, 171
judgment of, 170–72
language for, 167
margin of error and, 170
methods of, 171
numbers and, 168–69
online, 171
RealClearPolitics on, 168
samples for, 165
on 2016 election, 163–68
understanding of, 163–73
voter interests and, 172–73
voting impact from, 168–69
The Pollsters podcast, 167–68, 172
poll taxes
felons and unconstitutional, 50
Jim Crow laws on, 11, 13
of Native Americans, 17–18
popular vote
for Clinton, H., 179–80, 184
Electoral College and, xiv, 183, 185, 190–91
Prohibition, women support of, 34–35
Project VOTE!, 81
proper news story, news media and, 153–55
property ownership, Adams, J., on, 5
protests. *See* parades and protests
Purvis, Robert, 6

racial gerrymandering
causes and effects of, 131
League of Women Voters on, 135
majority-minority districts and, 130
Supreme Court in illegality of, 130, 133
Rainey, Joseph, 9
rally to end gerrymandering, *127*
ranked-choice voting, 64–66
*Ratf**ked* (Daley), xiv, 132
RBG (documentary), 142
Reagan, Ronald, 52, 54
RealClearPolitics, on polling, 168
Reconstruction Acts (1867–1868), 8–9
REDMAP, of FairVote, 132–33
Reed, Ralph, 92
regional and local sources, for news media trust, 151
registration, x
ACLU on voter required training and, 48
AVR increase of, 67–68
confirmation of, 195
of high school population, 76, 80
methods to increase, xiii
peer-to-peer, 80, 82–83
state election laws on, 47
TurboVote on, 119
as voting problem, 46–47
of youth vote, 80–83
Remember the Ladies (Dodson), 31
Remnick, David, 51
Revels, Hiram, 9
Rhimes, Shonda, 73
The Right to Vote (Keyssar), 7
RISE. *See* Ross Initiative in Sports for Equality
Roberts, Cokie, 43
Roberts, John
on gerrymandering, 122
as Supreme Court judge, 54–55, 136
Roberts, Rebecca Boggs, 37
Roberts rules, 55–57
robocalls, 111
Rock the Vote youth movement, 75
Rodriguez, Modesto, 53
Roosevelt, Franklin D., ix, x, xv–xvi
on political equality, xi
Ross, Stephen, 107–8
Ross Initiative in Sports for Equality (RISE), 107–8
Rucho v. Common Cause (2019), 136, 139–40
Reuters, 150–51

Ruffalo, Mark, 74
Ruth, Janice, 37

Sader, Jerome, 88
same-day registration (SDR), 69
Sargent, Aaron, 32
Sargent, Ellen Clark, 32
Schmidt, Michael, 153–55
Schwarzenegger, Arnold, 136, 138
SDR. *See* same-day registration
Section 2, of VRA, voter ID and, 56
Section 5, of VRA, 52–53
  Hyde on, 54
  Shelby County v. Holder on, 55–56
Seneca Falls Convention, women's suffrage and, 27–28
Senecal, Jeanette, 63
  as non-news source, 160–61
Shahidi, Yara, xiv, 73, 74, 77–79
  on news media, 146
Shaw, Anna Howard, 37–38, 40
Sheehy, Marie, 99
Shelby County v. Holder (2013), 57, 78
  on VRA Section 5, 55–56
Sherman, Ellen Ewing, 35
Shires, Amanda, 96
Silver, Nate, 164
Sisneros, Art, 175–77
slavery
  abolition of, 29
  Electoral College history impacted by, 182
"Slay the Dragon" campaign, of Fahey, 140–43
Slay the Dragon (documentary), 142–43
Smith, Marissa, 96, 97–98
Smith, Paul, 139
Snyder Act (1924), 17
social media, 146
  for youth vote, 74–75
social-pressure communications, 111–12
Sotomayor, Sonia, 137
sport organizations, RISE on voting in, 107–8
Stanton, Elizabeth Cady, 27, 28, 35
state election laws
  on absentee ballot, 47
  Bryce on, 47
  on early voting, 47
  on felons and voting, 48–50
  on gerrymandering, 138–39
  on Native American vote, 17
  on registration, 47
  variability of, 47
  for voter ID, 57
states
  Electoral College and small, 187–89
  electors choice by, 179
  population and Electoral College, 179
  swing, 183, 187–89
  2016 presidential election polling errors, 166
Stelter, Brian, 147, 150–51
Stewart, Celina, 60–61
Stone, Lucy, 30
Student Nonviolent Coordinating Committee, 13–14
suffragette, 120–21
Sugayan, Andaya, 80–81
Sullivan, Timothy, 41
Supreme Court. *See also specific judges*
  Brown v. Board of Education, 13
  Common Cause v. Lewis, 139–40
  Dred Scott case, 7, 13
  on electors removal, 180
  Husted v. A. Philip Randolph Institute, 59
  Lawrence v. Texas, x, 139
  on partisan gerrymandering, 134–38
  Plessy v. Ferguson, 12–13
  on racial gerrymandering illegality, 130, 133
  Rucho v. Common Cause, 136, 139--40
  Shelby County v. Holder, 55–57, 78
Susan B. Anthony Amendment. *See* Nineteenth Amendment
Swing Left, 91–92
swing states, 183, 187–89

Tapper, Jake, 159
team, youth vote importance of, 89–92
Teele, Dawn Langan, 33–34
Terrell, Mary Church, 30–31
thank-you notes, as voter encouragement tactic, 113
Tilden, Samuel, 185, 186–87
Time magazine, 151–52
Toobin, Jeffrey, 159
Totenberg, Nina, 136
Towles, Barry, 97
Tran, Natalie, 117
Traugott, Anderson, *109*
Trujillo, Michael, 18
Trujillo, Miguel, 17–18
Trump, Donald, 107
 2016 Electoral College votes, 166, 177, 184, 188
trust, of news media, 153
 book sources of, 152
 FactCheck.org, 151
 magazine sources for, 151–52
 national networks for, 150
 podcasts, 152
 Politico, 151
 regional and local sources, 151
 of Reuters and Associated Press, 150–51
Truth, Sojourner, 28, 30–31
TurboVote, 119
TurboVote.org, 193
Twelfth Amendment, 180
2012 reelection campaign, of Obama, B., 46
2016 presidential election, 164–65, 167–68
 AVR registration and, 67
 Clinton, H., and, 163, 183–84, 188
 Electoral College vote in, 166, 177, 179, 183–84, 188
 immigrants and, 116
 inconsistent responders for, 166
 swing states and, 183
 Trump Electoral College votes, 166, 177, 184, 188
 voters late change in, 166
 voting lack in, xiv, 21, 46, 75
 youth vote in, 74, 91–92
2016 presidential election polling, 163–64
 inaccuracy of, 165–68
 state polls error rate on, 166
2020 election, gerrymandering and, 143–44
Twenty-Sixth Amendment, on youth vote, 22

unbiased sources, in news media, 149
underserved contact, for youth vote, 86–88
union, voting motivation from, xii

Veep (TV show), 178
Vietor, Tommy, 81, 89
volunteerism, 91
vote-at-home, 70
vote-by-mail, 70
VOTE411, of League of Women Voters, 63, 148, 160, 195
 VOTE411.org, 193
Vote.org, 193
voter encouragement tactics
 business voting initiatives steps, 102–3
 Green and Gerber on, 111–12
 social-pressure communications, 111–12
 thank-you notes as, 113
voter fraud, 61
 confusion as reason for, 62–63
 Kobach on, 62
voter ID, 55
 conflict and confusion on, 57–58
 Election Day state requirements of, 57
 First Amendment and, 139
 FiveThirtyEight on laws for, 58
 Native Americans and, 51, 96
 VRA Section 2 and, 56
voter illusions, 119–20

voter interests, polling and, 172–73
voter rolls
exact match review for, 60
monitoring of, 58–61
voter rolls, purging of, 51
Brennan Center study on, 59
election notice response failure and, 59–60
election skipping and, 59–60
League of Women Voters on, 62–63
voters
ACLU on required training and registration of, 48
2016 presidential election late change by, 166
Voters Not Politicians, of Fahey, 140–43
voter suppression, xiii
Ho on, 56–57
of immigrants, 53
League of Women Voters fight on, 58–59
Marziani on, 51
Stewart on, 60–61
types of, 51–52
voter turnout, 111–12
of low-income Americans, 86
voting history impact on, x
Vote Save America effort, 90–91
#VoteTogether parties, 118, 119, 196
voting
of African Americans, 5–16
Anthony fine for, 32
checklist for, 193–97
as civic duty, 101–2
empowerment of, 74
Gronke on factors for, 110–11
of immigrants, 19–21
informing friends for, 199
low-income Americans turnout for, 86
of Native Americans, 16–19
personal awakenings for, 105–6
polling impact on, 168–69
procedures for, 63–64
state election laws on period of, 47
2016 presidential election lack of, xiv, 21, 46, 75
U.S. opportunities for, xiii–xiv
voter eligibility and lack of, xiv
voting, early days
of free black men, 5–6
property ownership for, 5
of white-men-only rule, 5
of white property-owning men, 5
voting problems and solutions, 45
AVR and, 66–69, 99
felons and, 48–50
holdups in, 47–48
ranked-choice voting, 64–66
registration and, 46–47
Roberts rules, 55–57
SDR, 69
vote procedures, 63–64
voter fraud, 61–63
voter ID conflict and confusion, 57–58
voter rolls monitoring, 58–61
voter suppression, 51–52
voting percentages, 45–46
VRA renewals, 52–55
voting rights. *See also* women's suffrage
for African Americans, x, 43
for Asian Americans, x
of men of all races, 6, 9, 30
for Native Americans, x
Voting Rights Act (VRA) (1965), 3
African American women and, 43
Bloody Sunday and, 13–14
extension for youth vote, 22
Fifteenth Amendment and, 15
gerrymandering and, 130
impact of, 15–16
on minority vote, 6–7
Section 2 of, 56
Section 5 of, 52–53
voting age lowering by extension of, 22
Voting Rights Act (VRA) (1965) renewal, 55

Carter on, 53–54
immigrants and, 53
Nixon on, 53
Reagan on, 52, 54
#Voting Squad, of When We All Vote, 89–91
VRA. *See* Voting Rights Act

Wallace, George, 186
Wasserman, David, 128, 132, 133
wave election, 133, 144
Wells-Barnett, Ida B., 31, 38
The West Wing (TV show), 177–78
We Vote Next for 2020, 77
We Vote Next Summit, *73*, 77–79
When We All Vote, of Obama, M., 88
get-out-the-vote language of, 115
#VotingSquad of, 89–91
White House protesters, March 1965, 3
white-men-only rule, 5
white property-owning men, 5
Wilson, Woodrow, 36, 40–41
women. *See also* African American women
African American rights support by, 30–32
discord, in women's suffrage, 30, 34–35
disenfranchisement of, 29
education vote by, 26
neglect, women's suffrage and, 25–27
New Jersey vote for, 26
Nineteenth Amendment suffrage for, 6
political offices of, 24
Prohibition support by, 34–35
women's suffrage use by, 43–44
Women's Loyal National League, 28–29
Women's Rights Convention (1851), 28
women's suffrage, x, 6, 23, 23–24
African American women and, 30–31, 34–35
in Britain, 35–36
discrimination, 29–32
Douglass on, 28, 30, 31
movement for, 27–29
parades and protests, 35–38
political parties and, 33–34
pros and cons of, 38–39
Seneca Falls Convention and, 27–28
Wilson and, 36, 40–41
women discord in, 30, 34–35
women neglect and, 25–27
women use of, 43–44
Woodhull, Victoria, 32
World War II (WWII), Native American veterans and, 17
Wyman, Kim, 70

Young, Andrew, 53
youth vote, xiii
apathy and, 93
CIRCLE on, 75–76
draft age impact on, 21
on Election Day, 83
for gun reform, 76
high school registration for, 76, 80
language for, 77–79, 115–16
learned behavior for, 84–86
movements for, 75–76
personal importance for, 93–94
positive peer pressure and, 88–89
registration of, 80–83
Shahidi and, xiv, *73*, 74, 77–79, 146
social media for, 74–75
team importance, 89–92
in 2016 presidential election, 74, 91–92
Twenty-Sixth Amendment on, 22
underserved contact for, 86–88
VRA extension for, 22

ABOUT THE AUTHOR

Erin Geiger Smith is a journalist who has written for the *Wall Street Journal* and the *New York Times*, among many other leading publications. She also worked at Reuters covering legal news. Before becoming a reporter, she was a lawyer practicing commercial litigation in New York and Texas. Erin grew up in Liberty, Texas, and graduated from the University of Texas at Austin, the University of Texas School of Law, and the Columbia University Graduate School of Journalism. She lives in New York City with her husband, Bryan, and their son, Reed.